ARITHMETIC
for
ENTERTAINMENT

Yakov Perelman

For general information on our products and services, please contact us on prodinnova@mail. com

Printed in United States.

ISBN : 9782917260388

10 9 8 7 6 5 4 3 2 1

ARITHMETIC
for
ENTERTAINMENT

Yakov Perelman

Contents

Foreword

This book is different from other books of similar content,[1] not in term of the material it contains but in term of the way this material is processed and presented. It does not only study the arithmetic topics studied in school, but also analyzes how these ones can be used in various other areas and in solving real life problems. Additionally, it does not try to turn enjoyable problems into tedious tasks that are often fruitless for serious work. It avoids difficult problems and selects only the material that is accessible for the majority of the readers.

Although the book is recommended for readers who are already familiar with the basic notions of arithmetic, it can also be used by beginners who are willing to discover this field of mathematics.

---

1 Among them the famous collection of E.I. Ignatieff "In the realm of wit" which contains three books and covers almost all the "classic" material in arithmetic entertainment.

Foreword

CHAPTER I

The Old and the New Numbers and Numeral Systems

Mysterious signs

In the early days of the Russian revolution (March 1917) residents of Petrograd were quite puzzled and even alarmed by some mysterious signs that appeared at the doors of several apartments. Rumor attributed these signs to various reasons. Those that I had seen were in the form of exclamation marks that alternated with crosses (these ones looked like the ones placed on top of tombs). In the general opinion these were not good a good thing as they frightened bewildered citizens. Sinister rumors went by the city. People talked about a band of freebooters that used to signs to mark the houses of their future victims.

The "Commissar of Petrograd," trying to calm the population, claimed that the "The inquiry revealed that the mysterious signs that appeared on the doors of ordinary citizens in the form of crosses, letters, shapes, have been made by provocateurs and German spies". He invited residents to erase and destroy these signs.

There are mysterious and sinister exclamation marks and crosses at the door of my house and my neighbors' houses. Some experience in resolving intricate puzzles helped me, however, to solve this simple and not-so-complex cryptography secret. I hastened to share my "discovery" with my fellow citizens by placing the following note in the newspapers:[1]

Mysterious signs
"In connection with the mysterious signs that appeared on the walls of several houses in Petrograd, it would be

---

[1] The evening edition of the newspaper "Stock exchange Sheets" from March 16, 1917

Chapter I

useful to clarify the meaning of one category of such signs, which, despite their ominous style, are of very innocent origin. I'm talking about the signs of this type:

$$†\text{!!}\quad ††\text{!!!!!}\quad †††\text{!!!}$$

Similar signs are seen in many houses on the back of the stairs or at the doors of the houses. Typically, signs of this type are present on all current house doors, and within one house two identical signs may be observed. Their grim mark naturally inspires anxiety to tenants. Meanwhile, their meaning is quite innocent, and can be resolved easily, if we compare them with the numbers of the corresponding owners. For example, the above signs are found at the doors of houses number 12, 25 and 33 respectively:

$$†\text{!!}\quad ††\text{!!!!!}\quad †††\text{!!!}$$
$$12\qquad 25\qquad 33$$

It is not difficult to guess that crosses mean tens and sticks mean units. This rule allowed solving all the occurrences I have seen without exception. This numbering rule was made by Chinese workers[2] who do not use our numbers."

I suppose these numbers were there before the revolution However, they only caught the attention of anxious citizens now. Mysterious characters of the same typeface, (but with oblique crosses instead of straight ones), were

2 They had spent a lot of time in St. Petersburg. Later I learned that the Chinese character for 10 is just the above specified form of a cross. The Chinese do not use our "Arabic" numbers.

Yakov Perelman

found in the houses where the workers were Russian peasants who came from villages. It is not hard to figure out the true authors of these cryptographic characters and we should not expect that their "artful" designations of houses should cause such a stir.

Ancients' numbering

Where did Petrograd workers find this simple method for numbers representation: crosses for tens, sticks for units? Obviously, they do not come up with these signs by themselves. They brought them out of their villages, where such signs were already in use since a long time and were known to everyone even the illiterate peasants in the most remote and obscure corners of Russia.

No doubt, these signs go back to ancient times and were widely used not only in Russian but in other regions of the world. It is worth mentioning their striking closeness to Chinese symbols and to the simplified Roman numbers. Indeed, in Roman numerals, sticks are used to represent units, and oblique crosses are used to represent tens.

Curiously, this was a popular numbering that was once used and legal in Russia: tax collectors used a similar, but more developed system, and recorded taxes in notebooks using similar signs. "The collector - we read in the old "code of laws" –accepting from households the amount he requested, should (or is clerk) record that amount in a notebook against the name of the household. The amount is recorded using figures and signs. The signs that are used should be identical everywhere for easy reference, namely:

Ten Ruble . . . . , . . . □

Ruble . . . . ○

Ten Kopeck . . . . ✕

Kopeck . . . . |

Quarter . . . . —

Using the above conventions, we can represent for example twenty-eight rubles fifty seven cents three quarters as follow:

Elsewhere in the same volume of "Code of Laws", there is again a reference to the mandatory use of the popular numeric designations. A special sign is used for thousand of rubles in the form of a six-pointed star with a cross in it, and another one is used for a hundred rubles in the form of a wheel with eight spokes. But symbols for the ruble and ten cents have been left unchanged from the previous law.

Here is the text of the law on these "tributary signs":

"On each of the receipt issued by the Collector, in exchange for the introduced tribute, the amount should be included in both words and using special signs made by rubles and kopecks, so it could be assured a fair reading[3]." Signs used in receipt mean:

---

3 This is a confirmation that these signs were used widely among the population.

Yakov Perelman

(star)	Thousand Ruble
(wheel)	Hundred Ruble
□	Ten Ruble
×	Ruble
| | | | | | | | | |	Ten Kopeck
|	Kopeck

For example, depicting 1232 rubles and 4 kopecks leads to the following figure:

As you can see, our Arabic and Roman figures are not the only way to represent numbers. In the old days, and even now in villages, we have used other systems of written figure, vaguely similar to the Roman but very far from the Arabic figures. But we have not yet identified all the ways used to represent numbers in our days: Merchants for example, have their own secret signs that they use

Chapter I

to represent numbers. We will talk about them more in detail in the next section.

A trade secret

On objects that are brought in local markets - and often in the stores, especially in the provinces - you've probably noticed strange designation letters like *ul, me,* etc.

These are not real prices. Instead, they represent a solution to allow the seller to remember the pricing information without the buyer being able to know this information. At a glance at these letters, the seller immediately gets their meaning and the premium he could make if he offers a certain price to the buyer. This notation is very simple - if you only know the associated "key". Usually the trader selects a word made of 10 different letters. Examples include *Yaroslavel, mirolyubet, Miralyubov, etc.* The first letter of the word is 1, the second is 2, the third is 3, etc. the last letter of the word denotes zero. With these letters, the trader is able to write down a price on the product which is kept strictly secret as the buyer does not know the "key" of the trader system. If for example, the trader selected the word

mirolyubet 1234567890

the price of 4 rubles 75 kopecks will be designated as follow: *o ul*
the sign *m lt* means 1 ruble 50 kopecks, etc.

Sometimes, products prices are indicated using numbers, but under the price there is also a letter designation. For example:

Yakov Perelman

$$\frac{\textit{3} \text{ rubs } \textit{50} \text{ kops}}{\text{bt}}$$

This means that the proposed price is 3 rubles 50 kopecks. Using the key *"mirolyubet"* means that the seller would not be able to offer a discount of than 80 kopecks.

The objective of the seller is to keep his "key" strictly confidential and protected. But if you buy in the same store a few objects and compare the prices of these objects with the corresponding letters, it is easy to guess the meaning of the letters. Particularly, it is easy to find out the secret word for cheap products where it is difficult to get significant discounts, and consequently, the first digit of the paid amount corresponds to the first letter in the notation. By guessing some of the letters, it is possible to find the other ones, and thus, it would be possible to crack the "key".

Let's say for example that you bought few products at a store, and paid for the first one 14 rubles, for the second – 12, for the third – 17 rubes. On these items you find the designations *mo, mi, mu.*

Clearly the letter m represents 1, and so m is the first letter of the word. Then you try to match the letters o, i, and u, with different digits. It is easy to notice that the word *mirolyubet* is a match. Now you can verify your guess by purchasing another product and checking its designation.

Chapter I

Arithmetic over a breakfast

After what has been said, it is easy to understand that numbers can be represented not only with digits, but also using any other signs or objects - pencils, pens, rulers, erasers, etc.: You only need to set a rule that ascribes each object to a particular digit or number value.

You can even, for the sake of curiosity, with the help of objects represent operations on numbers - addition, subtraction, multiplication, and division. For example, a number of operations on numbers can be completed using tableware (see Figure below) as representations. Fork, spoon, knife, jug, kettle, and dish are objects that can be used to substitute numbers.

Analyze the above group of knives, forks, dishes, etc., and try to guess the correspondence between digits and objects.

At first glance, this task seems very difficult: it is necessary to solve multiple equations, as once did Frenchman Jean-François Champollion. But your task is much easier: you do know that the numbers here, though represented as forks, knives, spoons, etc., are written in the decimal system, i.e., you know, that when a plate is standing in the second position (from the right), it represents a certain digit of dozens, similarly when the plate is in the extreme right, it represents the same digit of units, and when it is in the third position, it represents the same digit of hundreds. In summary, you know that the location of these items has a specific meaning that remains valid in the arithmetic operations carried the numbers they are used in. All of this can greatly facilitate the resolution of the proposed problem.

Yakov Perelman

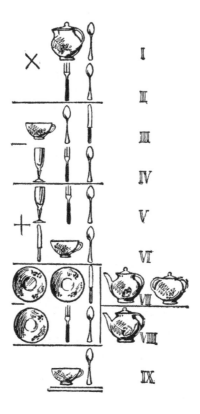

Here's how you delve the values of the arranged artifacts. Considering the first three series in our figure, you can see that when a "spoon" is multiplied by the "spoon", it gives a "knife." And from the following series you can see that when a "spoon" is subtracted from a "knife", we obtain a "spoon" or that "spoon" + "spoon" = "knife". What is the digit that gives the same result when it is doubled or squared? This can only be 2 because $2 \times 2 = 2 + 2 = 2$. Thus we know that the "spoon" = 2, and consequently, the "knife" = 4.

Let's continue. Which digit is indicated by the fork? Let's try to solve this, look to the first three series where the

22

fork is involved in multiplication, and to the series III, IV and V, where the same fork appears in the operation of subtraction.

From the subtraction series, you can see that subtracting in tens digit a "fork" from the "spoon", we get as a result a "fork", i.e. by subtracting a "fork" from 2, we get a "fork". This leads to two possible cases: either "fork" = 1 and then we have 2 − 1 = 1; or we have a "fork" = 6, and then subtracting 6 from 12 (as there is a "cup" in the higher digit) we get a 6.

Which digit should be chosen: 1 or 6? Let's test whether 6 suitable for the other operations. Pay attention to the addition of the V and VI series: the "fork" (i.e. 6) + "cup" = "plate". This means that a "cup" should be less than 4 (because the rows of VII and VIII show that "plate" − "fork" = "cup"). But the "cup" cannot be equal to 2, as this digit is already taken by the "spoon". Additionally, a "cup" cannot be 1 – otherwise the subtraction of series IV from III could not give a three-digit number in the V series. Finally the "cup" cannot be a 3 and here is why: If the "cup" was a 3, then the "glass" (see series IV and V) should designate 1; because 1 + 1 = 2, i.e., "glass"+"glass" = "cup" − 1 which was used during the subtraction of the tens. Similarly a "glass" cannot be equal to 1 because the "plate" in the VII series will mean in one case digit 5 ("glass" + "knife") and in the other case digit 6 ("fork" + "cup"), which is impossible. Hence, it is impossible for a "fork" to be equal to 6. Consequently a "fork" is equal to 1.

Knowing that a "fork" is equal to 1, it is now possible to have more confidence and to go more quickly. Observing the subtraction in III and IV series, we can see that the "cup" is either 6 or 8. Digit 8 has to be rejected, because it would lead to "glass" = 4 and we already know that digit 4 denotes a "knife". So the "cup" represents digit 6 and hence the "glass" represents digit "3".

What is the digit indicated by the "jug" in series I? It is easy to respond, as we know a product (624 from series III) and one of the two factors (12 from series II). Dividing 624 by 12, we get 52. Consequently, the "jug" is equal to 5.

The meaning of the "plate" is determined very simply: in series VII "plate" = "fork" + "cup" = "glass" + "knife", i.e., "plate" = 1 + 6 = 3 + 4 = 7.

Now we need to find out the digital value of the teapot and the sugar bowl in the VII series. As the digits 1, 2, 3, 4, 5, 6 and 7 are already taken the only possible remaining digits are 8, 9 and 0. Substitute the digits to the objects in division shown in the last three rows,[4] you will obtain the equation below (the letters y and x indicate the "teapot" and the "sugar bowl" respectively):

$$774 : yx = y$$
$$\underline{712}$$
$$62$$

Number 712, is the product of two unknown numbers "yx" and "y", which, of course, cannot be zero nor end with zero: hence, neither x, nor y are zeros. There remain two assumptions: $y = 8$ and $x = 9$, or vice versa $y = 9$ and $x = 8$. But multiplying 98 by 8, does not lead to 712, hence, the "teapot" stands for 8 and the "sugar bowl" stands for 9, (this leads to: $89 \times 8 = 712$). We have unraveled the

4 The arrangement of the numbers here is the one used and accepted in England and America.

hieroglyphic inscription from the objects on the dining table:

Jug = 5
spoon = 2
fork = 1
cup = 6
glass = 3
teapot = 8
sugar bowl = 9
plate = 7

The whole series of arithmetic operations represented by the objects now acquires a meaning:

$$\times\begin{array}{r} 52 \\ 12 \end{array}$$

$$-\begin{array}{r} 624 \\ 312 \end{array}$$

$$+\begin{array}{r} 312 \\ 462 \end{array}$$

$$-\begin{array}{r} 774:89 = 8 \\ 712 \end{array}$$

$$62$$

Decimal bookshelves

The decimal notation is wittily used even in areas where at first sight it is not expected - namely, in the distribution of books in the library.

Yakov Perelman

Usually, when you want to indicate the librarian the number of the book that interests you, you ask him to provide with a catalog of the books available and then you look for the desired number. This is because each library has its own numeral system which bookers numbers are defined. However, there is a system of books distribution in which any book has the same number in all libraries. This system can be seen as the decimal system of books classification.

This system – which is not adopted yet everywhere unfortunately – is extremely useful and is not very complex. Its essence lies in the fact that every branch of knowledge is associated with a certain digit, so that the number of the book gives information about its subject in the general system of knowledge.

Books are primarily broken down into ten broad classes, designated by the digits from 0 to 9:
0. Works of general nature.
1. Philosophy.
2. Religion.
3. Social Sciences.
4. Philology.
5. Physics, Mathematics and Natural Science.
6. Applied Science.
7. Fine Arts.
8. Literature.
9. History and Geography.

When numbering a book in this system, the first digit to the left indicates the class (among the ten above) to which the book belongs. Thus, each book on philosophy has a number starting with 1, in mathematics - with 5, in safety - 6. Conversely, if the number of the book begins, for example, 4, then we are not opening the book to know its subject we can say that it is a work of the field of linguistics. Further, each of the ten classes listed above are divided into 10 major departments also defined using digits. These digits are placed in second place of the book number. So if we consider the 5th class (Physics and mathematics and natural sciences), the following department are defined:

50. General writings on physics and mathematics and natural sciences.
51. Mathematics.
52. Astronomy. Surveying.
53. Physics. Mechanic.
54. Chemistry.
55. Geology. Paleontology.
56. General Geography.
57. Biology. Anthropology.
58. Botany.
59. Zoology.

Likewise, the other classes are broken down into departments. For example, in the class of Applied Sciences (6) we have the following departments: Department of Medicine indicated by digit 1 after 6, i.e., number 61, agriculture - 63, housekeeping - 64, trade and communication routes - 65, industry and technology - 66, etc.

Then there is a third digit in the numeral system that defines the content even more in detail. This digit points to the exact sub-division of the department to which the book belongs. For example, in the Department of Mathematics (51), a third digit of 1 indicates that the book refers to arithmetic; digit 2 - algebra, etc. Therefore, all the books on arithmetic have the following first three digits

511 in their numbers, on algebra – 512, geometry - 513, etc. Similarly, Department of Physics (53) is divided into 10 sub-divisions: books on electricity designated number 537 on optics - 535, etc.

The numeral system includes further subdivision using the fourth digit, etc.

In a library arranged with such system, finding the right book is simplified to the extreme. For example, if you are interested in geometry, you just go to the shelves where the numbers start with 5, find that the shelves where the stored books start with number 51... and then restrict yourself to shelves where book numbers start with 513...; similarly, when searching for books on cooperation, you look in the shelf where the book numbers start with 331... without checking the catalog and without bothering anyone with questions.

No matter how vast is the library it would not run out of numbers when considering this numbering of books. Conversely, the lack of books in a section does not prevent the use of this system: Indeed, some sections will remain unused.

Our favorite numbers

You have probably noticed that each one of us has a favorite among the numbers for which he nourishes a special predilection. For example, we are very fond of "rounded numbers", i.e., these ending with 0 or 5. This predilection for certain favorite numbers that are preferred to others is inherent in the human nature and is much deeper than it is usually thought. This predilection can be seen with all Europeans and their ancestors (such as the ancient Romans) and with populations in other parts of the world.

At every census, there is an unusually excessive abundance of people whose age ends with 5 or 0. They are much more numerous than they should be. The reason is, of course, that people do not remember exactly how old they are, so when asked about their age, they involuntarily "round" the number of years. A similar prevalence of "rounded" ages can be observed on the gravestones of ancient Romans.

This numeric bias goes very far. German psychologist Professor K. Marbach calculated the frequencies of ages written on Roman tombstones, and compared these frequencies with these of ages obtained from a census in the U.S. state of Alabama, which is the home of predominantly analphabet blacks. He found a surprising result: Ancient Romans and modern blacks in the US exhibited the same frequencies of ages. Namely, the more frequent unit digits are in this order: 0, 5, 8, 2, 3, 7, 6, 4, 9 and 1.

But that's not all. To investigate the numeric preferences of modern Europeans, Marbach carried out the following experiment: He requested several people to estimate "with their eyes", the length of a finger-long strip of paper in millimeters, and to write down the answers. He then counted the frequencies of different answers and curiously found that the most frequently used digits were in the same above order: 0, 5, 8, 2, 3, 7, 6, 4, 9 and 1.

It is no accident that people, so remote and different from each other - both anthropologically and geographically – exhibit the same feelings for certain numbers, i.e., a clear predilection for "round" numbers ending with 0 or 5, and a significant hostility toward non- rounded numbers (These terminating with 1, 9, 4, and 6).

You can perform a similar experiment yourself and check that you will obtain similar results: You can offer to a large audience to name any number between 1 and 10, then between 11 and 20, then 21 and 30, then 31 and 40, and finally 41 and 50. You will find that most of the answers will end with 5 with the remaining digits appearing less frequently. In other words, you will be able to measure the above predilection exhibited by people to some digits.

People predilection for numbers ending with 5 or 0 is, without a doubt, in direct connection with the decimal base used by our numeral system, and ultimately the number of fingers in our hands. But the approximations made by humans in order to get a 5 and 10 remain unexplained. Unfortunately, these ones do not come without a penalty.

Most people do not realize that our addiction to rounded numbers is costing us quite dearly. Commodity prices in retail always gravitate toward these rounded numbers, and in case of a non rounded number, it is always moved to the next higher rounded number. This process is thus done at the expense of the buyer, not the seller. The total amount that the country is paying for the pleasure of having rounded prices is quite impressive. Long before the last war, someone went through the trouble of roughly estimating this amount. It turned out that the population of Russia annually overpays in the form of the difference between rounded and non-rounded commodity prices more than 30 million gold rubles.

CHAPTER II

Stumbling Blocks in the Pythagorean Table

Stumbling blocks in the Pythagorean table

Most of us forgot about the time when we had to learn the multiplication table and gradually overcame it line by line. However, some may recall that not all the rows of the table generate the same difficulty. Some rows are assimilated very quickly, almost from the first reading - for example $5 \times 5 = 25$, $8 \times 2 = 16$. Others are much more difficult: first, they are remembered, but soon disappear again from memory, so we had to go back to the table several times before they are firmly imprinted in our memory. Recall how soon you managed to learn by heart $7 \times 8 = 56$. At least, for many it was one of the most difficult rows of the table.

Meanwhile, it is necessary to master and know the entire table by heart: the modern way of multiplying and dividing multi-digit numbers is based on the solid mastering of one-digit numbers multiplications, i.e., on the knowledge of the Pythagorean table by heart.

In an effort to facilitate this work, educational psychology experts focused their attention on the most difficult parts of the multiplication table and subjected them to detailed study. The results were quite interesting. It turned out that the main stumbling blocks in the table are recurrent among people, namely the lines listed here below:

$$7 \times 8 = 56$$
$$9 \times 7 = 63$$
$$9 \times 8 = 72$$
$$7 \times 6 = 42$$
$$9 \times 6 = 54$$

Chapter II

Among several hundred surveyed adults and children, the majority indicated that these five lines of multiplication were the most difficult in the entire table. Especially, they unanimously referred to line $8 \times 7 = 56$.

The next lines in the multiplication table in term of difficulty are:

$$8 \times 6$$
$$8 \times 8$$
$$7 \times 6$$
$$8 \times 4$$
$$7 \times 4$$
$$7 \times 5$$
$$7 \times 3$$
$$5 \times 4$$
$$8 \times 5$$
$$6 \times 4$$

The researchers also looked to the most difficult columns in the Pythagorean table. Using the same careful questioning, they requested a sample of people which of the 10 columns in the multiplication table is the most difficult to learn. The answers were unanimous: Namely, and in order, the columns of multiplication by 7, 8, 9, and then 6. In contrast, the easiest ones were unanimously – and as expected – 2, 3, 5, and 4.

The results of these psychological studies[1] are likely to coincide with the conclusions of the personal experience of most readers. With no doubt, we all agree that the cases of multiplication by 7, 8 and 9 have been and remain the most difficult to digest and that the most difficult

---

1 Max Duhring ("Zeitschr. F. Päd. Psychol.", 1912).

Yakov Perelman

of all are the rows: 8 × 7, 9 × 7, 9 × 8, 7 × 6 and 9 × 6. The debate is only limited to their degree of difficulty. Even adults, who victoriously overcome all arithmetic difficulties, sometimes stammer on these cases when they have to calculate in a hurry or are tired. In these cases, not trusting their memory, they try to check the result using a workaround or ask for confirmation from others: "Seven eight - fifty six?"

Obviously, these difficulties are not random, since they occur with extreme regularity. How to explain them?

There are several reasons for this phenomenon and all of them are rooted in the unconscious techniques we normally use to memorize numbers. With multiplications that are considered easy, we have some supplementary support and help (although usually we do not know or suspect it). For example, when multiplying by 2, we unconsciously replace this operation by one that is more familiar to us in the form of an addition: 4 x 2 = 4 + 4. There is also the help of the memorization consonance: "five times five - twenty-five", "six times six - thirty six", "six times eight - forty eight." Rhymed lines are always easier to remember, especially at a young age.

There are many other circumstances that facilitate the memorization of numbers in the Pythagorean table. These circumstances if listed would lead to a long list and these ones have not been established beyond dispute anyway. Why the row 9 x 9 = 81 is easier to remember than 7 x 8 or 8 x 9? Probably the characteristic pattern of number 81 (a curved eight and a straight-line one) helps. Another example is number 5 and the fact that all the products resulting from multiplying by this number ends with a 5 or a 0. Other cases are easy to remember because of their frequent use in everyday life (e.g. 4 x 7 - four weeks).

Special difficulties were noticed with those five multiplication cases which took most of the votes during the surveys. This may be related to the fact these are no specific circumstances (such as the above) that apply to them and that facilitate their memorization:

$$8 \times 7 = 56$$
$$9 \times 7 = 63$$
$$8 \times 9 = 72$$
$$7 \times 6 = 42$$
$$9 \times 6 = 54$$

These numbers are rarely found in everyday life, and they do not have any specific easily memorable visual pattern. The fact that these rows all include four different but similar numbers (8, 7, 6, and 9) also makes memorization difficult. Finally, close results, such as 56 and 54, are easily mixed and a clear distinction between them requires significant efforts. These subtle features are the root cause that turns these rows of the multiplication table into constant stumbling blocks for anyone learning this table.

Multiplication using fingers

To facilitate the assimilation of the multiplication table, we can resort to our hands' fingers: Using them as a simplified calculating machine. We can use them to automatically find the results of several multiplications. Obviously we still need to know the results of some multiplications by heart (such as 5 x 5).

Using fingers is a vintage way to do multiplications and

Yakov Perelman

it is still used by people in Siberia, Ukraine and in some remote corners of Livonia. However, it would be very useful to acquaint all students with this multiplication technique. Suppose that you want to calculate 7 × 9. Bend on one hand as many fingers as the difference between 7 and 5, and on the other – as many fingers as the difference between 9 and 5 - in short bend over the excess of 5 on each side. We will have the following:

	Bent	Straight
On one hand	2	3
	+	×
On the other hand	4	1

Now, sum up the number of bent fingers (2 + 4 = 6), add a zero to the right of the result (60), and add to this number the product of the straight fingers (3 x 1 = 3). You will get 63.

Another example: Let's try to calculate the result of 6 × 8:

	Bent	Straight
On one hand	1	4
	+	×
On the other hand	3	2
Result	**4**	**8**

Proceed the same way, you will obtain 48. You can see that the simplicity of this technique can hardly hamper even the young mathematician if this one is familiar with the first part of the Pythagorean table. This technique already allows to effortlessly get hold of the more difficult part of the table.

Chapter II

This multiplication technique using fingers is described in Mignitsky's "Arithmetic" using the following terms:

"Fingers can be used to perform multiplications in the same way as the multiplication tables. Let suppose that you want to calculate 7 x 7. You bend two fingers on one hand and two fingers on the other one, as 2 is the difference between 7 and full 5 fingers. Sum the numbers associated with the bent fingers and append the result with a zero. You obtain 40. Multiply the numbers associated with the straight fingers, you will obtain 3 x 3 = 9. Then, sum the two numbers 40 + 9 = 49. This is the result of the multiplication."

What is the secret of this technique? We will understand this if we consider the general case. A small excursion into "general arithmetic", i.e. algebra, will convince us that this technique should give the correct results in all the cases from 6 × 6 to 10 × 10. Every number that is greater than 5, can be written in the form of 5 + a, 5 +b, 5+ c, etc.

In all these expressions, the letters a, b, and c are the excess amount over number 5. Consequently, the product of two numbers that are greater than 5 can be written using this notation as follow:

$$(5 + a) \times (5 + b)$$

or as in algebra multiplication does not require writing a specific sign:

$$(5 + a)(5 + b).$$

And what do we do when we multiply using the fingers?

Yakov Perelman

We bend a fingers on one hand and b fingers on the other, while leaving the other fingers straight, i.e., on one hand $(5 - a)$ and on the other $(5 - b)$ fingers are left straight. Then we add $a + b$ and add a zero, i.e. the number 10 $(a + b)$. To this is added the product of the straight fingers, i.e. $(5 - a)$ $(5 - b)$.

Therefore, the total result is $10(a + b) + (5 - a)(5 - b)$.

If we perform the multiplication of the content of the two brackets we obtain $25 - 5a - 5b + ab$. To this, we add $10a + 10b$, which leads to a total of $25 + 5a5b + ab$, which is exactly $(5 + a)(5 + b)$.

In short, all the actions on the fingers can be represented in the general form as follow:

	Bent	Straight
On one hand	a	$5 - a$
On the other hand	b	$5 - b$
Result	$10 (a +b) + (5 - a) (5 - b).$	

And we already know that this expression is equal to $(5 + a)(5 + b)$.

As we have said at the beginning of this section the multiplication using the fingers can be performed for numbers up to 15 × 15. How can this be done? It is somewhat different from the multiplication of up to 10 × 10.

Suppose you want to multiply 12 × 14. Fold the excess factors above 10 on each hand (and not above 5 as before), i.e. on one hand fold two fingers, and on the other fold four. Add the numbers of folded fingers on each hand (2 +

4) and multiply the result by 10, and then add to the result the product of the folded fingers (4 x 2), and then add to the result 100. We have: $12 \times 14 = 100 + (2 + 4) \times 10 + 4 \times 2 = 168$.

Another example: 11×13:

		Bent
On one hand	. . . .	1
On the other hand	. . . .	3
Result	. . . .	$100 + 40 + 3 = 143.$

What is the secret behind this technique? Let's go back to Algebra again. In all cases, this multiplication can be generally represented as follow:

$$(10 + a) \times (10 + b)$$

Where a and b are numbers that are smaller than 5, and represent the bent fingers. Performing multiplication using the general rule, we obtain:

$$(10 + a)(10 + b) = 100 + 10(a + b) + ab.$$

From this formula, we can deduce how the technique is built.

Curiously enough, the product 10×10 can be found using fingers with both techniques. Indeed, with the first, we have:

	Bent	Straight
On one hand	5	0
On the other hand	5	0
Result	$(5 + 5)\ 10 + 0 + 0 = 100.$	

And with the second technique:

	Bent
On one hand	0
On the other hand	0
Result	$100 + 10\ (0 + 0) + 0 \times 0 = 100.$

There is also a technique for the multiplication of numbers from 15 × 15 to 20 × 20 using fingers, but this technique is too complex. Every calculating machine is good when treated at its just value. Our natural machine is no exception to this rule.

Mechanical multiplication by 9

We describe another –simple but interesting technique that is used when multiplying numbers by 9. Suppose that you have to multiply 7 × 9. You can proceed as follow: Extend your hands in front of you on a table, and then bend the 7th finger from the left. Consequently, you have six straight fingers on the left side and three straight fingers on the right side. These two digits give the desired product: 63. In order to multiply 5 by 9, bend the 5th finger. You will have 4 fingers on the left and 5 on the right. The product of the multiplication is thus 45.

The reader can try to provide an explanation for this technique.

Yakov Perelman

CHAPTER III
Descendants of the Ancient Abacus

Chekhov's problem

Everyone probably remembers the famous arithmetic problem that has so confused and embarrassed the seventh-grader Ziberov from Chekhov's story "Tutor":

"A merchant bought 138 yards of black and blue cloth for 540 rubles. The question is, how many yards he bought from each, if the blue cloth cost 5 rubles a yard, and black cloth costs 3 rubles?"

With subtle humor Chekhov describes how the seventh graders' tutor and his disciple Ziberov helplessly worked on this problem until they were later rescued by his brother Peter and their father Udodov:

Ziberov repeated the problem and immediately, without saying a word started dividing 540 by 138.
- Why do you divide? Wait a minute! However, so ... go ahead. What is reminder? There can be a reminder. Let me divide!

Ziberov performed the division, found a remainder 3, and quickly blurred.
- Strange ... - he thought, ruffling his hair and blushing. - How should it be done. Ahem! ... This is an indeterminate equation, not an arithmetic one.

The tutor looked at the answers and saw 75 and 63.
- Hm!... Strange ... Add 5 to 3, and then divide 540 by 8? Is that it? No, not that!
- Decide on! said Peter.
- Well, what do you think? The problem is surely a breeze,

said Udodov

- What a fool you, brother! said Peter.

The tutor took the pencil and began to solve. He hesitated, turning red and pale.

- This problem is, in fact, algebraic, - he says. - It can be solved with an x and a y. However, it could be solved. I've shared ... See? Or, that's what. Solving this problem seems difficult... Think...

Peter smiled maliciously. Udodov smiled too. Both realized the embarrassment of the tutor and the student.

- Even without algebra, the problem can be solved – said Peter while Udodov chose another path: Holding out his hand to the abacus. - Here, you can see it...

He moved the stones, and turned 75 and 63.

This scene with this problem makes us laugh at the embarrassment of tutor, but in turn, raise three new problems, namely:

1. How can the problem be solved algebraically as suggested by the tutor?
2. How the problem was resolved by Peter?
3. How the father used the abacus to find the desired numbers?

While the first two questions can be probably easily answered – at least by many readers of this book, the third question is not that simple. But let's look to three questions in order:

(1) The tutor considered solving the problem "with an

Yakov Perelman

x and a y", being sure that the problem - "is, in fact, algebraic." And indeed, he would have easily resolved it if he resorted to the help of a system of equations. The two equations in this case are: $x + y = 38$, $5x + 3y = 540$, where x and y are the numbers of yards of blue and black cloth respectively.

(2) However, the problem can also be easily solved arithmetically. If you had faced it, it wouldn't have stalemated you. You would start with the assumption that all the bought cloth was blue. If the entire batch of 138 yards was blue, the merchant would have paid $5 \times 138 = 690$ rubles for it. This is $690 - 540 = 150$ rubles more than what was paid in reality. The difference of 150 rubles indicates that the merchant has also bought cheaper black cloth at 3 rubles per yard. Cheap cloth was so much that the difference of cost was 150 rubles while being 2 rubles for each yard: obviously, the number of yards of black cloth is determined by dividing 150 by 2. Thus, it is easy to get the answer: 75 yards of black cloth and the difference $138 - 75 = 63$ of blue cloth.

(3) Now the third question is how Udodov was able to resolve the problem. The story is very succinct about this. It only says: "He moved the stones, and turned 75 and 63." However, which "stones" did he move? In other words, what is the way he used to solve this problem using an abacus?

Fortunately, solving the problem using an abacus is similar to solving it arithmetically using paper. But the abacus makes the resolution much easier. This is because of the benefits that our Russian abacus provides anyone who knows how to handle it. Obviously, as a retired provincial secretary, Udodov was experienced with the use of an abacus. Using experience, he knows how to look for the desired numbers without resorting to the use of "an x and a y" like the tutor. Here are the steps needed to resolve the problem using an abacus:

First of all he multiplies 138 by 5. In order to do so, he acts according to the rules in the abacus. He first multiplies 10 by 138 (i.e., just moved 138 one wire above) and then halved this number (i.e. separated the stones into two haves on each wire). If the number of seeds on this wire is odd, then to get out of trouble, he needs simply to transfer one stone from the tens into the lower wire. In our example, were we have to divide 1380 by 2, the result is that the top wire would not contain any stones, the next one would contain 6, followed by 9, while the bottom wire would remain empty. Consequently, we know that the result of the division is 690. Obviously, for an experienced user, these operations are done automatically.

Next Udodov had to subtract 540 from 690. How this is done in the abacus - we all know that.

Finally, the resulting difference (150) has to be split in half: Udodov cancelled the top stone and provided 10 to the second. He then left 7 on each side of the second and gave an additional 10 to the last wire. These ones were split on two parts. The result of the division was 75.

The above operations are illustrated on the abacus as follow:

Yakov Perelman

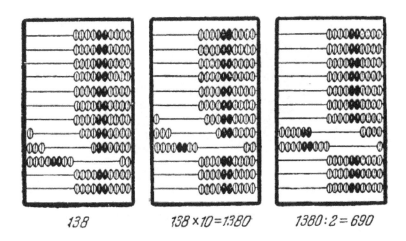

138   138 × 10 = 1.380   1380 : 2 = 690

Obviously, all these simple actions are performed on the abacus much faster than what is described here.

Russian abacus

There are many useful things that we cannot appreciate simply because they are constantly used and have turned into very ordinary objects in our household.

One of such objects that are not given their due appreciation is, without doubt, our Russian abacus. This one is a mere modification of the famous "abacus", or "counting board" of our remote ancestors. Ancient civilizations – the Egyptians, Greeks, Romans – have used basic counting devices that significantly resemble our Russian "abacus".[1]

1 It was the board (table), on which bands were drawn and which allowed the movement of special checkers which played the role of seeds in our Abacus. This board was called the Greek abacus. The Roman abacus was in the form of a copper board with grooves on which buttons were moved. The Peruvian abacus was in the form of belts with knots on them. This one was particularly widespread among the original inhabitants of South America, but was most certainly used in Europe as well.

A Russian abacus

In the Middle Ages and until the XVI century, such devices were widely used in Europe. But nowadays, enhanced abacuses have been preserved only in Russia and China. The West does not want these devices anymore. You will not find them in any store in Europe. Perhaps, it is because of this that we do not appreciate this counting device as it should be. Look at it as some sort of primitive calculating machine.

Meanwhile, we had the right to be proud of our "abacus", which thanks to its marvelous simplicity, can achieve impressive results and compete in some respects, even with complex, costlier calculating machines from Western countries. In skilled hands, this simple instrument can perform miracles. Thus, it is not surprising that when foreigners get acquainted with this device, they appreciate it much more than us.

There was one specialist who was in charge of one the largest Russian companies selling calculators. He told me that he used the Russian abacus to amaze a countless number of foreigners who brought him complex foreign calculators. He even arranged for a contest between two human counters, one worked on an expensive foreign calculator while the other worked on a Russian abacus. It often happened that the latter – who, truth to be said, was a master of his craft – bested the former in both speed and precision. It also happened that when a foreigner was struck by the rapidity of the people using the abacus, immediately gave up, folded his complex machine back into its suitcase, and lost any hope of selling a single unit in Russia.

- Why do you need expensive calculating machines, if you calculate so skillfully using your cheap machines! - Often said representatives of foreign firms.

Indeed, foreign calculating machines are hundreds of times more expensive than Russian abacus but they produce more actions and allow more operations. Nevertheless, in many areas such as the addition and the subtraction for example, the abacus can safely compete with complex foreign mechanisms. Even multiplication and division are significantly accelerated in skillful hands - if you know the special techniques for performing these actions.

Let's look at some of these techniques.

Calculating the outcomes of multiplications

Here are a few techniques that can be used by anyone (who is familiar with additions) to quickly perform multiplications:

Multiplication by 2 and 3 is simply done using the summations (adding the number to itself once or twice).

To multiply by 4, first multiply by 2, and then add the result to it itself.

To multiply by 5, you can proceed as follows: Add a zero (i.e. multiply by 10) and then take the half of the resulting number (i.e. divide by 2).

To multiply by 6, you can multiply by 5 and then add the original number to the result of the multiplication.

To multiply by 7, you can multiply by 10 and then take out 3 times the original number.

To multiply by 8, you can you can multiply by 10 and then take out the double of the original number.

Similarly, to multiply by 9, you can you can multiply by 10 and then take out the original number.

The reader will now probably have figured out by himself how to proceed when multiplying by a number that is greater than 10 and what kind of replacements will be more appropriate:

To multiply by 11, you must multiply by 10 and add the original number to the result (i.e. use the fact that 11 = 10 + 1). To multiply by 12, multiply by 10 and then add double the original number to the result (i.e. use the fact that 12 = 10 + 2). To multiply by 13, multiply by 10 and then add three times the original number to the results (i.e. use the fact that 13 = 10 + 3), etc.

Here are a few techniques that you can use for the first hundred factors:

Yakov Perelman

$$20 = 2 \times 10$$
$$22 = 2 \times 11$$
$$25 = (100/2)/2$$
$$26 = 25 + 1$$
$$27 = 30 - 3$$
$$32 = 22 + 10$$
$$42 = 22 + 20$$
$$43 = 33 + 10$$
$$45 = 50 - 5$$
$$63 = 33 + 30, \text{etc.}$$

It is easy to see, among other things, that multiplications by numbers like 22, 33, 44, 55, etc., are easy to calculate. Therefore, we should aim to break down the multipliers in a way to introduce such numbers. We can resort to the same techniques when multiplying by numbers that are greater than 100. If the techniques are tedious, we can always use the general rule and multiply the number by each digit of the multiplier and then doing the addition of the results.

Calculating the outcomes of divisions

Performing divisions is more difficult than multiplications. You will have to remember a set of special techniques, sometimes quite complex, to be able to quickly complete divisions. Here we will only explain the techniques for the first dozen numbers (except for number 7, for which the method is extremely complex).

We already know how to divide by 2; the associated technique is very simple.

The division by 3 is slightly more complex: If we try to replace this division by a multiplication, we will have to multiply by the infinite periodic fraction 3.3333 ... (as we know that 0.333 ... = 1/3). We can easily multiply numbers by 3 and we can also easily divide them by 10 (this one is too easy): You can repeatedly add 3 times the original number to the result while each time dividing by 10. After some short practice, this method of division by three, which was at first glance complex, would look easy.

Obviously, in order to divide by 4, we simply have to divide twice by 2.

In order to divide by 5, we simply have to divide by 10 and then double the result.

In order to divide by 6, we need to proceed in two steps: first we need to divide by 2 and then we need to divide the result by 3.

Dividing by 7 is too complex to have a simple practical rule.

In order to divide by 8, we need to proceed in three steps: first divide by 2, then divide the result by 2 again, and then divide the result by 2 another time.

The division by 9 is very interesting. It is based on the fact that 1/9 = 0.1111.... It is clear that, instead of dividing by 9, you can repeatedly add 0.1 to 0.01, 0.001... of the original number to the result.[2]

As you can see, divisions by 2, 10, and 5 are easy, consequently, it is also easy to divide by their multiples

2 This technique is useful for mental division by 9.

Yakov Perelman

(4, 8, 16, 20, 25, 40, 50, 75, 80, and 100). The reader can easily devise rules for the division by these numbers.

Echoes of antiquity

Some vestiges from the remote ancestors of our Russian abacus have survived and are still used in the common language and the popular customs. Few suspects, for example, that by tying a knot in a scarf for "memorization", we repeat what was once a great way for our ancestors to "record" an account. A cord with knots represented an accounting device that was basically similar to our abacus and, most certainly, was connected to them by the generality of origin: it is a "rope abacus."

A rope abacus

The same abacus is linked to some currently widespread words like "bank" and "check". "Bank" in German means bench. What is common between a "bank" or a financial institution in the modern sense of the word, and a bench? It turns out that this is not a mere coincidence. Abacuses in the form of benches were widespread in German business in the XV-XVI centuries. Each money-changing store or banking office was characterized by the presence of "Counting bench" - and of course, the bench has become synonymous with the bank.

The abacus is indirectly related to the word "check". This word is of English origin and is derived from the verb "checker" (or chequer). The relationship comes from a "checkered" leather cloth that was called abaca and that English merchants carried with them in XVI-XVII centuries to perform calculations. It could be easily carried out in a pocket and deployed on a table for the purpose of calculus. These cloths developed over time and morphed into calculating devices. The word "check" as we know it today comes from these "chekered" clothes.

Curiously, the expression "to stay with the beans" which we now use to refer to a person who lost all his money is of very ancient origin, and goes back to the time when all cash transactions - including payments between people – were calculated using an abacus. Beans were used on a bench and played the role of the seeds in our abacus.[3] People who lost their money, stayed with some beans that expressed the sum of their losses. This expression survived and is still used even if beans are no more used in calculus.

3 Some used pebbles, others used beans as mentioned by Campanella in "The City of the Sun" (1602).

Yakov Perelman

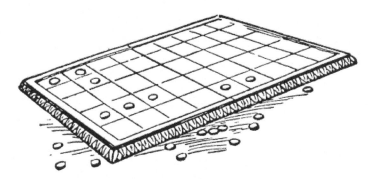

A pebbles' abacus

Chapter III

CHAPTER IV
Some History

"The difficult thing - division"

When we light a match, we don't think about it, but in fact we should use that moment to think about the efforts that were required from our ancestors to make a fire, and this was not in the very distant past. Similarly, very few suspect how it was difficult for our ancestors to perform the four arithmetic operations. Fortunately, the new calculus techniques have made it easier and more convenient to complete such operations and quickly obtain the desired results. Our ancestors used techniques much more cumbersome and slow. And if a student of the XX century could be transported four, even just three centuries back, he would have struck our ancestors by the speed and infallibility of his arithmetical calculations.

Rumors about him would spread through to the surrounding schools and monasteries. He would outshine the most skilful calculators of that era. From all over Europe, people would come to learn new tricks from the new great calculus master...especially operations such as multiplication and division which were complex and difficult for our ancestors.

However, returning back to the past is not possible. Consequently, our ancestors have to content themselves with a whole dozen different ways of multiplication and division, which were similar in their complexity. They were all confusing, hard to remember, and difficult to use by average person. Every calculus teacher tried to promote his favorite technique, and every "division master" (there were such specialists) tried to invent their own techniques.

Chapter IV

62

There were multiple multiplication techniques known under different names: "the chess", "the bending", "in part or in the gap", "the cross", "the grating", "the backwards technique", "the triangle", "the cup or the bowl", "the diamond" and others.[1] Similarly, there were multiple division techniques with no less fanciful names. They vied with each other in complexity.

These were digested with great difficulty and only after a long practice. It was even acknowledged that mastering the art of fast error-free multiplication and division of numbers requires a special natural talent and exceptional ability which ordinary people did not have.

"The difficult thing... division" said an old Latin proverb, and justifiably division uses laborious, tedious methods that involved multiple manipulations. It does not matter that these methods have often quite playful names. Hidden under these funny names, there are generally long and tedious series of intricate manipulations. In the XVI century, the shortest and most convenient way to perform a division was called "boat" or "galley". Famous Italian Mathematician of the time Niccolo Tartaglia wrote about this technique in its detailed arithmetic tutorial as follows:

"The second technique of division is called Boat or Galley in Venice[2], owing to the vague resemblance of the figure obtained in the process of the division to a boat or a galley. In all cases, the figure looks indeed really beautiful; the

1 These techniques are described in Niccolo Tartaglia book "Arithmetic". Our modern way of multiplication was called "the chess" at that time.

2 Venice and some other cities of Italy in XIV-XVI centuries conducted extensive maritime trade. Consequently, arithmetic operations which were used in these places for commercial purposes, developed earlier than in others countries and the best works on arithmetic appeared in Venice. Many Italian terms related to commercial arithmetic are still used today.

Yakov Perelman

galley appears sometimes well decorated and equipped with all the accessories – the appearing numbers looks like a galley's feed, nose, mast, sail, and oars,..."

A galley in Venice

The ancient Italian mathematician strongly recommended this technique as "the most elegant, easiest, most reliable, and the most common for the division of all possible numbers" - I still do not dare to present it here, for fear that even the patient reader will get bored and close this book at this point without reading any further.

Meanwhile, this really tiresome technique was the best in that era. It was used in Russia until the middle of the XVIII century. In his "Arithmetic" book, Magnitsky described it as one of the six techniques that were available at the time (none of these ones was similar to the modern technique we use today) and has especially recommended it. In his

64

voluminous book (which contained 640 pages), Magnitsky described it as an extremely "arcade" technique, and did not use its "galley" or "boat" name.

To conclude, we show the reader a numeric "galley" using the example mentioned in Tartaglia's book:

	4 | 6	
88	1 | 3	08
0999	09	199
1660	19	0860
88876	0876	08877

09999480000001994800000019999
1666660000000866600000000866666

DIVIDEND — 888888000000088880000000888888 (88- QUOTIENT)

Divisor — 999990000000099900000000099999
999990000000099900000000099

Wise antic custom

Arriving after tiring efforts to the desired end of arithmetic operation, our ancestors considered it necessary to have some means to check the correctness of the produced results. Additionally, because the used calculus methods were cumbersome, they created a natural distrust of their results. Indeed, it is easier to get lost on a long tortuous path than on a straight road that uses modern techniques. This lead naturally to the ancient custom of checking each performed arithmetic operation. It is recommended to follow this custom as it does not hurt us.

Yakov Perelman

One favorite trick to achieve this goal is the so-called test 9. It is a very elegant technique that is useful for the reader to know. It is often described in modern arithmetic textbooks, but for some reason, it is not frequently used in practice, which, however, does not reduce its merits.

The verification is based on the following rule (known as the "balances rule"): The reminder of the division of a sum by any number is the sum of the reminders from dividing each term by the same number. On the other hand, it is also known that when a number is divided by 9, the obtained reminder is the same reminder obtained when the sum of its digits is divided by 9. For example, when dividing 758 by 9, we obtain a reminder of 2, and when the sum of its digits (5 + 7 + 8) is divided by 9, we obtain the same reminder (2).

Using these two properties, we can create a verification method called the "balances rule" or "the rule of nines."

Suppose you want to check the correctness of the addition illustrated here below:

```
   38932 . . . . . . . . . . . 7
    1096 . . . . . . . . . . . 7
 4710043 . . . . . . . . . . . 1
  589106 . . . . . . . . . . . 2
 5339177 . . . . . . . . . . . 8
```

You need to mentally sum the digits of each term, then in the resulting numbers, also sum up the digits. Repeat the process of adding up the digits until at the end you get only one single digit. The results of this reduction process are recorded next to each numbers as shown in the figure above. Now, if you sum these numbers and repeat the same process with them, you will obtain 8. This is the same number that is obtained if we do the same simplifications with the sum of the original numbers (5339177). Indeed 5 + 3 + 3 + 9 + 1 + 7 + 7 will lead after simplification to 8 (this is the reminder modulo 9).

To check the subtraction, we can proceed in the same way, if we take the sum and reduce it and then subtract the resulting numbers, we would obtain the same result from reducing the difference. For example:

$$
\begin{array}{r}
6913 \ldots\ldots\ldots 1 \\
-\ 2587 \ldots\ldots\ldots 4 \\
\hline
4326 \qquad 6
\end{array}
$$

4 + 6 = 10, i.e., 1.

Additionally, it is not complicated to check multiplications, as shown in the following examples:

$$8713 \ldots \ldots 1$$
$$\times$$
$$\underline{264 \ldots \ldots 3}$$
$$34852 \qquad 3$$
$$52278$$
$$\underline{17426}$$
$$2300232 \ldots \ldots 3$$

If such verification detects an error, then, to determine where the error is, you can check each of the multiplications using the same process to verify that there is no error with each of them. You should also check, using the same process, that there are no errors with the additions. Such testing saves time and labor. It is useful only when considering large multi-digit numbers. With small numbers, it is easier to redo the operation.

The use of this method with division requires little explanation. If you have a case of division without remainder, the check is performed as for multiplication: the original number is regarded as the result of the multiplication of the divider by the result of the multiplication. In the case of division with remainder, we should use the fact that the original number is equal to the divisor x quotient + residue. For example:

$$16201387 : 4457 = 3635; \quad \text{reminder} \quad 192$$

sum of digits 1 2 8 3

$$2 \times 8 + 3 = 19; 1 + 9 = 10; 1 + 0 = 1.$$

In "Arithmetic", Magnitsky offered the following representations for the verification process:

For multiplications:

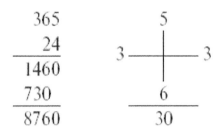

$$
\begin{array}{r}
365 \\
24 \\
\hline
1460 \\
730 \\
\hline
8760
\end{array}
$$

For divisions:

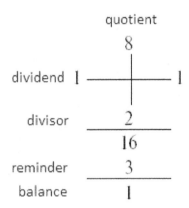

It is without a doubt, a verification method that does not leave much to be desired in terms of speed and convenience. However, you cannot say the same about its reliability: Errors may elude this method. Indeed, when a number has its digits permutated, the reminder of its division by 9 would not change. Worst, sometimes, when some digits are replaced by others, the replacement passes

Yakov Perelman

undetected by such verification method. It is also possible to hide from the control of this method using extra nines and zeros, as they do not affect the reminders of division by 9. Consequently, you have to be prudent when using this method for checking.

Our ancestors were aware of the limitations of this method, so they used it but still did an additional check – often with the help of 7. This method still uses the same "balances rule", but is not as convenient as the nines, because when dividing by 7, you have to complete fully the division to find the reminder (and it is easy to make error in the verification process itself). However, two checks (using nines and sevens) are sufficiently reliable to verify an arithmetic operation. Indeed, if an error eludes one check, it is highly likely it will be caught by the other one. Note, however, that there some errors that are able to elude both checks. This occurs when the difference between the obtained (erroneous) result and the real one is a multiple of number 7 x 9 = 63. Fortunately, in usual calculations, where mistakes are generally 1 or 2 units, it is possible to restrict ourselves to the nines check. The sevens check is overly burdensome. A check is good only when it is convenient.

"Russian" method of multiplication

In some places, our peasants sometimes use, out of necessity, a very ingenious way of multiplication of integers, that does not look like the way used in schools. This method is apparently inherited from the remotest antiquity. What makes the method interesting is the fact that using it, you can complete multiplications without the need of multiplication tables. The multiplication

70

of any two numbers is reduced to a series of successive operations dividing a factor into a half while doubling the other factor. Here is an example:

$$32 \times 13$$
$$16 \times 26$$
$$8 \times 52$$
$$2 \times 208$$
$$1 \times 416$$

The operations continue until the first factor is equal to one. Take the last second factor and you will obtain the desired result. The foundation of this method is obvious: the product does not change if one factor is halved while the other is doubled. Therefore, it is clear that as a result of these repetitive operations, we will obtain the desired product:

$$32 \times 13 = 416.$$

However, what should be done if during the process of halving the first factor, we obtain an odd number? It is easy to overcome this difficulty. It is necessary in the case of an odd number (while having the first factor) to put one unit of the corresponding second factor aside and continue the process with the next lower even number on the first factor side. The process of halving and doubling would continue until we reach one on the first factor side, but in order to obtain the correct product, it is necessary to add all the numbers that have been put aside to the result of the final line. In practice, it is possible to forget about all the rows containing even first factor and keep only the rows where the first factor is an odd number (including the last one with number 1 as a first factor). Add the second factors from these rows and you will obtain the product.

Yakov Perelman

Here is an example (the asterisk indicates that this line should discarded):

$$19 \times 17$$
$$9 \times 34$$
$$4 \times 68 \; *$$
$$2 \times 136 \; *$$
$$1 \times 272$$

Adding the second factors in the lines that should be kept, we obtain the correct result:

$$17$$
$$34$$
$$\underline{272}$$
$$323$$

It is not difficult to understand the full theoretical validity of this method, if we take into account that $19 \times 17 = (18 + 1) \; 17 = 18 \times 17 + 17 = 9 \times 34 + 17 = (8 + 1) \times 34 + 17 = 8 \times 34 + 34 + 17 = 4 \times 68 + 34 + 17 = 2 \times 136 + 34 + 17 = 272 + 34 + 17$.

As you see, you cannot deny the practicality of this popular technique of multiplication. "Knowledge" English scientific magazine dubbed it the "Russian peasant" way.

Country of the pyramids

It is very likely that this method came to us from ancient times and, moreover, from a far country - Egypt. We know very little about the arithmetic operations that were available to the inhabitants of the ancient land of the pyramids. But an intriguing monument was preserved - papyrus, on which were recorded arithmetic exercises students of the surveying schools of ancient Egypt. This so-called Rhind papyrus dates back to between 2,000 and 1,700 BC[3] and represents a copy of the more ancient manuscripts copied by a certain Aames.

Scribe[4] Aames found a "student's notebook" from that distant era. He carefully transcribed all the arithmetic exercises of future surveyors - along with their mistakes and the teacher's corrections - and gave to his list a solemn title which reached us in following incomplete form:

"Manual of how to achieve the knowledge of all dark things... all the secrets hidden in things. Compiled with the king of Upper and Lower Egypt, Ra-and-whisker, modeled on ancient writings from the time of King Ra-en-mat."

In this interesting document, which dates to about 40 centuries ago and more ancient times, we can find the following four examples of multiplication that are performed in a manner reminiscent of our Russian way (the points ahead of numbers indicate the number of units of the factor; the + sign indicate the number being subject of the addition):

3   A papyrus was spotted by British Egyptologist Henry Rhind, it was enclosed in a tin case. Unfolded, it was 10 yards long and 6 inches width. It is exhibited in the British Museum in London.

4   The title of "scribe" belonged to the third class of the Egyptian priests. Their responsibilities related to the construction of temples and their land ownership. Mathematical, astronomical and geographical knowledge were their primary specialty (V. Bobynin).

Yakov Perelman

(8 × 8)
. 8
... 16
.... 32
: : : : 64

(9 × 9)
. 9 +
.. 18
.... 36
: : : : 72 +

Total 81

(8 × 365)
. 365
.. 730
.... 1460
: : : : 2920

(7 × 2801)

. 2801 +
.. 5602 +
.... 11204 +

Total 19607

You can see from these examples that even thousands of years ago, the Egyptians have used a multiplication technique similar to our peasants, and that for some unknown reasons, these techniques have migrated from the ancient land of the pyramids to modern Russian villages. If ancient Egyptians were offered to multiply, for example, 19 × 17, they would have produced this operation as follows: They would write a series of successive doubling of number 17:

1	17 +
2	34 +
4	68
8	136
16	272 +

Chapter IV

And then they would sum the numbers that are marked above with a "+" sign, i.e., 17 + 34 + 272. They would have reached, obviously, the correct result: 17 + (2 × 17) + (16 × 17) = 19 × 17. It is easy to see that such a technique is essentially very similar to our "peasant" (replacement of the multiplication by a series of successive doublings).

It is difficult to say how Russian peasants managed to preserve such ancient way of multiplication and how they became known for it. British authors named it the "Russian peasant way". In Germany, some common people used this technique here and there and enjoyed it, but also called it "Russian."

It is interesting to get any available information from the readers about whether they apply this technique (or any other techniques) of multiplication in their respective regions, and this may provide glues about the origins of this one. Indeed, traditional mathematics should be given great attention with a focus on the techniques used by people for calculus and measurement. These ones should be considered the national monuments of mathematical creativity. They travelled the time from the depths of antiquity to our modern era.

Late historian of mathematics V.V. Bobynin even created a program for the gathering of mathematics antics. These ones included (1) Calculus techniques, (2) Methods of measure including weight, (3) surveying methods, (4) proverbs and riddles, etc.

CHAPTER V

Non-Decimal Numeral Systems

Mysterious autobiography

Let me start this chapter with a problem I came up with fifteen years ago for the readers of a magazine[1] when it was designed as "a prize problem." Here it is:

Mysterious autobiography

The autobiography of one eccentric mathematician started with the following lines:

"*I graduated from university at the age of 44. A year later, as a 100-year-old young man, I married a 34-year-old girl. There was a slight difference in age - 11 years - that contributed to the fact that we lived by common interests and dreams. After a few years, we were already a small family with 10 children. I was getting a salary of only 200 rubles a month, of which 1/10 had to be given to my sister, so we lived with the children using 130 rubles per month...*"

How can you explain the strange contradictions in the numbers of this passage?

The solution to the problem is suggested by the title of this chapter: non-decimal numeral systems - that the only explanation for the apparent inconsistency of the above numbers.

Let's look more into the idea. It is not difficult to guess what is depicted by the eccentric mathematician. Let's

---

1 "Nature and People" (and then was reprinted in E.I. Ignatieff's collection "In the realm of wit").

look at the first sentence "A year later (i.e. after the age of 44), as a 100-year-old young man..." So if the addition of one unit to number 44 results into 100, it means that digit 4 is the highest in the system (such as for example digit 9 in the decimal system), and hence we can deduce that the base 5 system is used. The Eccentric mathematician used the base 5 system through his entire biography, i.e., one in which the first digit represents the units (between 0 and 4) and the second digit does not represent the tens, instead it represents the fives, while the third digit represents the "twenty fives" instead of the hundreds, etc. Therefore, the number shown in the writing "44" does not mean a $4 \times 10 + 4$, as it is the case in the decimal system, instead it means $4 \times 5 + 4$, and that is equal to twenty four in the decimal system. Similarly, the number "100" in the mathematician autobiography has one unit in the third digit which represents the "twenty fives" in the base 5 system, so this number is equal to 25. The other numbers from the autobiography could be written as follow in the decimal system:

$$« 34» = 3 \times 5 + 4 = 19$$
$$« 11» = 5 + 1 = 6$$
$$«200» = 2 \times 25 = 50$$
$$« 10» = 5$$
$$« 1/_{10}» = 1/_5$$
$$«130» = 25 + 3 \times 5 = 40$$

Restoring the true meaning of the numbers, we don't see any contradiction at all:

I graduated from the university at the age of 24 years. A year

Yakov Perelman

later, as a 25-year-old young man, I married a 19-year-old girl.
The slight difference in age - only 6 years old - contributed to the
fact that we have lived by common interests and dreams. After
a few years, we were already a little family of 5 children. I was
getting a salary of 50 rubles, of which one fifth had to be given to
my sister, so we lived with the children for 40 rubles.

Is it difficult to represent numbers in other number
systems? Nothing could be easier. Let's suppose that you
want to represent number 119 in the base 5 system. Divide
119 by 5 to find out the quotient and the reminder:

$$119: 5 = 23, \text{ with a reminder of } 4.$$

Hence, the first digit will be 4. Further, 23 fives cannot fit
all in the second position. The highest digit we can get
in the second position in the 5-base system is 4. So let's
divide 23 by 5:

$$23: 5 = 4 \text{ with a reminder of } 3.$$

This shows that the number 3 will be located in the second
position ("fives") and number 4 will be located in the third
("twenty-fives") position. Thus 119 can be written as $25 + 4$
$\times 3 + 5 \times 4$, or "434" in the base 5 system.

The operations made can be summarized as follow:

119	5	
4	23	5
	3	4

In order to obtain the representation of the number in the

new base system, you simply need to write down the italic figures from right to left. You will immediately get the desired representation.

Here are some more examples.

(1) Represent number 47 in the ternary system:

$$
\begin{array}{c|c|c}
47 & 3 & \\
\hline
2 & 15 & 3 \\
\hline
 & 0 & 5
\end{array}
$$

Answer: "502." Verification: $5 \times 9 + 0 + 2 \times 3 = 47$.

(2) Represent number 200 in the septenary system:

$$
\begin{array}{c|c|c}
200 & 7 & \\
\hline
14 & 28 & 7 \\
\hline
60 & 0 & 4 \\
\hline
4 & &
\end{array}
$$

Answer: "404." Verification: $4 \times 49 \times 0 + 4 + 7 = 200$.

(3) Represent number 163 in the base 12 system:

$$
\begin{array}{c|c|c}
163 & 12 & \\
\hline
43 & 13 & 12 \\
\hline
7 & 1 & 1
\end{array}
$$

Answer: "117." Verification: $1 \times 144 + 1 \times 12 + 7 = 163$.

Yakov Perelman

I think that the reader is now capable of representing any number in any numeral system. The only issue that can arise is related to the fact that the 10 digits, as we know them, may not be enough to represent all the numbers. In fact, in a system with a base of more than ten (e.g., base 12 system) needs a representation for numbers 10 and 11. But it is not difficult to get out of this difficulty. This can be done by choosing any new figures, symbols or letters for these two numbers. For example, we can choose for number 10, the letter J which stands in the 10th position of the alphabet and similarly choosing letter K for number 11. Thus, the number 1579 is depicted in the base 12 system as follows:

$$
\begin{array}{r|l}
1579 & 12 \\
\underline{12} & \underline{131} \quad \begin{array}{r|l} & 12 \\ \hline 11 & 10 \end{array} \\
37 \\
\underline{19} \\
7
\end{array}
\qquad \text{«(10)(11)7»,} \quad \text{or} \quad 7.
$$

Verification: $10 \times 144 + 11 \times 12 + 7 = 1579$.

The simplest numeral system

Generally it is easy to realize that in each system the highest digit is obtained by subtracting one unit from the base of the system. For example, in the decimal system the highest digit is 9. In the base 6 system, the highest digit is 5. In the ternary system, the highest digit is 2, and in the 1base 5 system, the highest digit is 14, etc.

Chapter V

The simplest is the system, the fewer digits it requires. The decimal system requires 10 digits (0 and the 9 other digits), the base 5 system requires only 5 digits, the ternary system requires 3 digits (1, 2, and 0), and the binary system requires only 2 digits (1 and 0).

Now, the question is whether a 1-base system is really a system? Of course, it is a system in which a unit of a higher level is just another unit of the lower level, i.e. they are equal. We can probably call it the "single" system. This is the oldest "system" known to humans. It had been used by primitive men, who were making notches on the trees to record the number of countable items. There is a huge difference between this system and all the other numeral systems: A very important feature is missing from it: The so called packing value of digits. Indeed, in this "single" system, the character standing in the 3rd or the 5th position has the same meaning and is worth the same as the character in the 1st position. By contract, even in the binary system a unit in the 3rd position (to the right) is 4 times greater than a unit in the first position, and a unit in the 5th position is 16 times bigger than a unit in the 1st position. Therefore, this "single" system gives us very little benefits, as the representation of any number in it needs exactly the same number of characters as the observed items: In order to represent 100 items in this system, you would need 100 characters, while in the binary system you need only seven ("1100100") and in the base 5 system you need even less, just three ("400").

That is why the "single" system can hardly be called a "system". At least, it cannot be put at the same level with the others, as it is fundamentally different from them, not allowing any savings in the representation of the items.

Yakov Perelman

If we put this system aside, the simplest numeral system would be the binary system in which we use only two digits: 0 and 1. Using 0 and 1 it is possible to represent an infinite set of numbers! In practice, this system is not very useful as the numbers quickly become too long,[2] but theoretically it has all the rights to be the simplest. It has some interesting features that are exclusive to it. These features, among other things, can be used to perform a variety of spectacular mathematical tricks, which we will soon talk in detail in the chapter "Tricks without cheating."

Extraordinary arithmetic

We are so used to simple calculations that we do them automatically. However, these ones generate considerable stress if we try to apply them to numbers that are written in a system other than the decimal system. Try, for example, to perform the addition of the following two numbers written on the base 5 system:

$$+\frac{\text{«}4203\text{»}}{\text{«}2132\text{»}} \quad (\text{ In the base 5 system }).$$

Start the addition with the units, i.e., from the right: 3 + 2 equals 5, but we cannot write 5 because such digit does not exist in the base 5 system. This sum is equal to 1 higher level unit. Hence the sum 3 + 2 is equal to 10. At the second level we have 0 + 3 =3 in the addition to the unit obtained previously we obtain 4 units at the second level. At the third level, we have 2 + 1 = 3. At the fourth level, we have

---

2 But, as we shall see, the tables of addition and multiplication in such system are simplified to the extreme.

84

4 + 2 which is equal to 11 in the base 5 system. Thus the desired sum is 11340.

$$+ \quad \frac{\begin{array}{c}\text{«4203»}\\ \text{«2132»}\end{array}}{\text{«11340»}} \quad \text{(In the 5-base system)}.$$

The reader may verify this addition by moving the original terms to the decimal system, completing the addition, and then moving the resulting number back to the base 5 system.

Similarly, other operations such as subtractions and multiplications could be performed. Here below are some examples. These ones are given as exercises on which the reader can practice.

In the 5-base system

$$- \quad \frac{\begin{array}{c}\text{«2143»}\\ \text{«334»}\end{array}}{\text{«1304»}} \qquad \times \quad \frac{\begin{array}{c}\text{«213»}\\ \text{«3»}\end{array}}{\text{«1144»}} \qquad \times \quad \frac{\begin{array}{c}\text{«42»}\\ \text{«31»}\end{array}}{\begin{array}{c}\text{«42»}\\ \text{«231»}\end{array}}$$
$$\overline{\text{«2402»}}$$

In the ternary system

$$+ \quad \frac{\begin{array}{c}\text{«212»}\\ \text{«120»}\\ \text{«201»}\end{array}}{\text{«2010»}} \qquad \times \quad \frac{\begin{array}{c}\text{«122»}\\ \text{«20»}\end{array}}{\text{«10210»}}$$

«220»: 2 = «110»
«201» : 12 = «10»
(remainder «11»).

You can perform the above operations, by first, moving the original numbers to the decimal system, completing

Yakov Perelman

these operations in this new system, and then converting the result back to the original system. However, you can proceed using the following alternative: Prepare an "addition table" and a "multiplication table" for simple digits in the original system in which the inputs of the operations are supplied. For example, the addition table in the base 5 system is as follows:

0	1	2	3	4
1	2	3	4	10
2	3	4	10	11
3	4	10	11	12
4	10	11	12	13

With this table, we could add the numbers "4203" and "2132" written in the base 5 system, with much less straining efforts than the method employed earlier. Similarly, with this simplification, subtractions can be performed more easily.

Also, it is easy to prepare a multiplication table (also known as the "Pythagorean table") for the base 5 system:

1	2	3	4
2	4	11	13
3	11	14	22
4	13	22	31

With such table in your hand, you can easily complete multiplications (and divisions) in the base 5 system. You can easily verify the results shown in the above examples. For example, when multiplying:

$$\text{base 5 system} \begin{cases} \times \quad \text{«213»} \\ \quad\quad \text{«3»} \\ \hline \text{«1144»} \end{cases}$$

proceed as follow: From the table 3 times 3 is equal to "14". Write down 4 and keep 1 in mind. Then from the table 3 times 1 is equal to "3", adding the 1 from the units multiplication, we obtain 4. Write down 4. Then, again from the table 3 times 2 is equal to 11. Write down 11. We obtain the final result: "1144".

The smaller is the base system the less relevant are tables of addition and multiplication. For example, for the ternary system, the two tables are as follows:

0	1	2
1	2	10
2	10	11

Addition table for the ternary system

1	2
2	11

Multiplication table for the ternary system

They could be immediately remembered and used to perform actions. The smallest tables of addition and multiplication are obtained for the binary system:

Addition table for the binary system

0	1
1	10

Multiplication table for the binary system

$$1 \times 1 = 1$$

From such a simple "table", we can perform all the operations in the binary system. Multiplication in this system is in fact very simple: Multiplying by 0 leads to 0. Multiplying by 1 means leaving the other factor unchanged. Multiplying by "10", "100", "1000", etc. is as simple as adding the relevant number of zeros to the right. As for additions, you simply need to remember one result – that in the binary system 1 + 1 = 10. The binary system is the simplest system for basic arithmetic operations. This simplicity is unfortunately compensated by the extreme length of numbers of this system.

Let's consider, as an example, the following multiplication:

$$
\text{Binary System}
\begin{cases}
& \times \quad \ll 100101101 \gg \\
& \qquad\quad \ll 100101 \gg \\
\hline
& \qquad \ll 100101101 \gg \\
+ & \quad \ll 100101101 \gg \\
& \ll 100101101 \gg \\
\hline
& \ll 1010111011110001 \gg
\end{cases}
$$

In order to perform this operation, it is simply needed to rewrite the first number in the proper location: This requires less mental effort than the multiplication of the same number in the decimal system ($605 \times 37 = 22385$). If we have adopted the binary system, calculus would require far less mental effort (but unfortunately more paper and ink).

Odd or Even?

Without the representation of a number in the decimal system, it is difficult, of course, to guess whether it is odd or even. Consequently, if you see a number don't rush to say whether it is odd or even if you don't know in which system it is written. Let's consider number 16 and try to find whether it is odd or even.

If you know that this number is written in the decimal system, then without a doubt, you can say that it is an even number. But if it was written in some other system - can you be sure that it represents an even number?

The answer is no. If the base is, for example, seven, then "16" = 7 + 6 = 13, which is odd number. The same will happen for every odd base (because every odd number + 6 = odd number).

Hence the conclusion is that the familiar criterion for divisibility by two (the last digit is even) is suitable only for the decimal system. For other systems, this criterion is not always correct. Namely, it is valid only for systems with an even base value: base 6, base 8, etc. What is the criterion for divisibility by 2 for systems with an odd base? Here is a very brief criterion for odd bases: The sum of digits must be even. For example, the number "136" even in any system and even in odd base systems is an even number. Indeed, in this example we have an odd number + odd + even = even number.

With the same care, let consider the following problem: Is the number 25 divisible by 5? In the base 7 or base 8 numeral systems, 25 is not divisible by 5 (because "25" would be equal to 19 or 21 in these systems). Similarly, a well-known criterion for divisibility by 9 (the sum of the number digits should be divisible by 9) is correct only in the decimal system. In contrast, in the base 5 system this criterion is applicable to divisibility by four and, in the base 7 system, this same criterion is applicable to divisibility by 6. Thus, the number "323" in the base 5 system is divisible by 4, because 3 + 2 + 3 = 8, and the number "51" in the base 7 system is divisible by 6 (you can easily verify this by moving the numbers to the decimal system: you will get 88 and 36). Why do we have such criteria? The reader can figure the reason if he observes the criterion of divisibility by 9 and makes the same argument, suitably modified, for the base 5 and the base 7 systems.

Now, let's suppose we have the following equalities:

$$\left.\begin{array}{l} 121 : 11 = 11 \\ 144 : 12 = 12 \\ 21 \times 12 = 441 \end{array}\right\} \quad \text{These equalities are valid in one numbering system}$$

If you are familiar with the rudiments of algebra, you can easily find the base, and explain the properties of these equations.

A fraction without denominator

We are accustomed to the fact that fractions written down without the denominator are possible only in the decimal system. So at first glance it seems that writing directly without denominators $2/7$ or $1/7$ in another numeral system is impossible. But let not forget that writing a fraction without the denominator is also possible in other numeral systems. For example, what means the fraction "0.4" in the base 5 system? Of course, it means the $4/5$ fraction. "1.2" in the base 7 system means $12/7$. What the fraction "0.33" in the base 7 system means? Here the result is more complicated: $3/7 + 3/49 = 24/49$.

Here are a few examples of non-decimal fractions written without the denominator:
"2,121" in the ternary system is equal to $2 + 1/3 + 2/9 + 1/27$ $= 216/27$

Yakov Perelman

"1,011" in binary system is equal to $1 + 1/4 + 1/8 = 13/8$

"3,431" in the base 5 system is equal to $3 + 4/5 + 3/25 + 1/125$ $= 3116/125$

"2,555..." in the base 7 system is equal to $2 + 5/7 + 5/49 + 5/343 + ... = 17/6$

Chapter V

CHAPTER VI
Museum of Numeric Curiosities

Arithmetic curiosities

In the world of numbers, as in the world of living creatures, there are genuine rarities with exceptional properties. Such unusual properties could realistically be used to make some sort of a museum for numeric curiosities. In the galleries of such a museum, we would not only find a place for giant numbers which will be discussed in a separate chapter, but we would also find small numbers that possess extraordinary properties. Some of them are already able to attract interest and attention by their appearance, while others possess outlandish hidden properties. We invite the reader to visit with us this museum and get acquainted with some of these numeric pieces.

Let's pass without stopping past the first windows, as we are all familiar with the properties of the associated numbers. We already know the remarkable properties associated with number 2.

Not because it is the first even number, but because it is the most comfortable base (radix) we could use.

Also, let not be surprised when we meet number 5 here.

Besides 10, it is one of our favorites. It plays an important role in any "rounding", including the rounding of prices, which is costing us so dearly.

Similarly, we will not be surprised to find number 9.

$$\boxed{9}$$

Certainly not because of it being the symbol of permanence,[1] but because it is the number that facilitates arithmetic checks.

The real excitement starts when we reach the window behind which number 12 is exposed.

$$\boxed{12}$$

Why is number 12 a remarkable one? Obviously, this is the number of months in a year and the number of units in a dozen, but why is, in essence, a dozen so special? Not many people know that 12 is a very ancient number and almost beat other rivals (including number 10) to be used as a base (or radix) for our daily life calculations.

Civilizations of the ancient East - the Babylonians and their predecessors, the ancient original inhabitants of Mesopotamia – used 12 as a base for their daily life calculations. It is only due the powerful influence of India that we moved to the decimal system. Without that

1 Ancients (the followers of Pythagoras) believed 9 was the symbol of permanence, since the 10 first multiples of 9 retain the same sum of digits (9).

Yakov Perelman

influence, it is very likely that we would have stayed with the Babylonian base 12 system.

In some ways we are still using some element of the base 12 system, despite the victory of the decimal system. Our addiction to the dozen and to the Gross (which refers to 144 items or a dozen dozen), our division of the day into two dozen hours, the division of hours - 5 dozen minutes, and the minutes - the same number of seconds, our division of the circle into 30 dozen degrees, the division of the feet into 12 inches and many other units are eloquent examples of how great is the influence of this ancient system.

Do we have to be glad that in the struggle between the dozen and the ten, the latter finally won? Of course, our own ten fingers which can be considered as live counting machines are strong allies against the dozens. Without the fingers, we would have preferred the dozens instead of the tens. It is much easier to make payments in a base 12 system rather than in a decimal one. The reason for that is that number 10 is divisible only at 2 and 5, while 12 can be divided by 2, 3, 4 and 6. So instead of 2 divisors, we would have four. Other benefits of the base 12 system will become clearer if you take into consideration the fact that in this system, a number ending with a zero, is a multiple of 2, 3, 4, and 6: Think how convenient it would be to split a number into 1/2s, 1/3s, 1/4s, and 1/6s and the results would be integers.

Additionally, if a number expressed in the base 12 system ends with two zeros, then this one must be divisible by 144, and hence all the factors of 144, i.e., the following long series of numbers:

2, 3, 4, 6, 8, 9, 12, 16, 18, 24, 36, 48, 72, 144.

Chapter VI

So instead of having at least 8 divisors (2, 4, 5, 10, 20, 25, 50 and 100) for numbers ending with two zeros in the decimal system, you will have at least 14 divisors in the base 12 system. [2]

And while in the decimal system only the factions 1/2, 1/4, 1/5, 1/20 etc. are converted into decimal numbers, in the base 12 system, much more fractions can be converted into decimal numbers without the denominator. This includes the factions 1/2, 1/3, 1/4, 1/6, 1/8, 1/9, 1/12, 1/16, 1/18, 1/24, 1/36, 1/48, 1 / 72, 1/144, which are represented respectively as follow:

0.6, 0.4, 0.3, 0.2, 0.16, 0.14, 0.1: 0.09, 0.08, 0.06, 0.04, 0.03, 0, 02, 0.01.

With such obvious advantages to the base 12 system, it is not surprising that several mathematicians are calling for a full transition to this system. However, we have are now very used to the decimal system, and it is too late for such a reform.

You see, therefore, that a dozen has a long history behind

---

2 It would, however, be a great mistake to think that the divisibility of the number may depend on the system in which the value is depicted. If nuts contained in a bag, can be decomposed into five equal piles, this property will not depend on whether the number of nuts in the bag is expressed in a particular numeral system, written in words, or expressed in another different way. If a number, written in the base 12 system, is divisible by 6 or 72, then the same number written in the decimal system for example would have the same divisors. The only difference is that in a base 12 system, divisibility by 6 or 72 is easier to spot (for example number ending with one or two zeros). When people talk about the benefits of the base 12 system in term of a higher number of divisors, then keep in mind that this is due to our predilection for "round" numbers. Indeed, in practice, we would be able to find more rounded numbers in the base 12 system then in the decimal system.

Yakov Perelman

it and that number 12 has found itself in our numeric curiosities museum for a reason. But its neighbor – "baker's dozen", 13 is also in attendance.

$$\boxed{13}$$

Number 13 is featured here, not because it is a remarkable number, but because it is seen as a "scary" number by superstitious people despite the fact there is nothing special about it.

Number 365

This number is remarkable not only because it is the number of days in a year, but because, above all, its division by 7 gives a remainder of 1. This seemingly insignificant feature is very important in calendar calculations:

$$\boxed{365}$$

Because of this property, each simple (non-leap) year ends with the same day of the week with which it started. If, for example, New Year's Day was a Monday, then the last day of the year will be a Monday, and the next year will start from Tuesday. For this same reason - because of the reminder of the division of 365 by 7 is 1 - it would have been easy to change our calendar so that a given calendar date always fall on the same day of the week – for example if the 1st of May was Sunday in a given year, it would be a Sunday on each year. Thus, years would always start on the same day of the week and each date will be on the same day of the week from a year to another. In leap years which contain 366 days, we would have the first two days of the year billed as special holidays so that the above rule would be respected. This would significantly simplify the management of our calendar. Unfortunately, the number of days in a year is 365 and not 364. Thus, we have to deal with complex calculations in order to be able to find which day of the week is a given date in the past or the future.

Number 365 possesses also another feature that is not associated with the calendar but that is nevertheless very interesting. This one stems from the fact that:

$$365 = 10 \times 10 + 11 \times 11 + 12 \times 12.$$

Thus, 364 is the sum of the squares of three consecutive numbers, starting with ten:

$$10^2 + 11^2 + 12^2 = 100 + 121 + 144 = 365.$$

But that's not all: it is equal to the sum of the squares of numbers 13 and 14:

$$13^2 + 14^2 = 169 + 196 = 365.$$

There are not many numbers that possess this curious property.

Three nines

Another number that exhibits remarkable properties is built using three equal digits: 999

Yakov Perelman

999

This number is much more interesting than its inverted image - 666. This latter is the famous "beast number" or the Apocalypse number and it inspires huge fear in superstitious people, but its arithmetic properties do not stand out among the rest of the numbers.

A curious feature of number 999 appears when it is multiplied by any other three-digit number. You will get a six-digit number in which the first three digits to left reproduce the original number reduced by one unit, and the last three digits are obtained by calculating the difference between 999 and the number obtained from the first three digits. For example:

$$573 \times 999 = 572427 \quad \frac{\begin{array}{r} 572 \end{array}}{999}.$$

One has only to look at the following line to understand the origin of this feature:

$$573 \times 999 = 573 \times (1000 - 1) = \left\{ \frac{\begin{array}{r} 573000 \\ - \quad 573 \end{array}}{572427} \right. .$$

Chapter VI

From this feature, we can obtain a very simple method for "instant" multiplication of any three-digit number by 999:

$$847 \times 999 = 846153, \quad 509 \times 999 = 508491, \quad 981 \times 999 = 980019,$$
$$\text{etc.}$$

And since $999 = 111 \times 9 = 3 \times 3 \times 3 \times 37$, you can, with lightning speed, write a whole list of six-digit numbers that are multiples of 37. Obviously, someone who is not familiar with the properties of number 999 would not able to do so. In short, you can arrange small sessions of "instant multiplication and division" for the uninitiated and show you magician abilities like none else.

Glorified By Scheherazade

Next in line we have 1001 which was glorified by Scheherazade

You probably did not know that in the very title of the collection of magical Arabian Nights is also a kind of miracle that would capture the imagination of the fabulous Sultan beyond the other wonders of East, if he was interested in arithmetic curiosities.

Why is number 1001 so great? In appearance, it seems quite ordinary. It does not even belong to the category of so-called elected prime numbers: the sieve of Eratosthenes shows it is divisible by 7, 11 and 13 - three consecutive prime numbers whose product is exactly number 1001. But the fact that $1001 = 7 \times 11 \times 13$ is nothing magical. What is more remarkable is that fact when you consider any three-digit number and multiply it by 1001, you would obtain a number that consists of the original number written twice: for example, $873 \times 1001 = 873873$, $207 \times 1001 = 207,207$, etc. And although this is to be expected, since $873 \times 1001 = 1000 \times 873 + 873 = 878\,000 + 878 = 878,878$, it can be used to achieve completely unexpected results - at in front of an untrained person.

Namely, everybody who is not initiated to arithmetic mysteries can be the subject of the next trick. Have someone freely write a three digit number (that is unknown to you) on a piece of paper and then let him ascribe again the same number. He will get a six-digit number consisting of three repeating digits. Invite a friend to take this resulting number and divide it by 7. He will be surprised to find out that the number is divisible by 7. Request him to write down the result of the division on a paper and pass it to a second friend without you seeing it. Ask this second friend to divide the number by 11. Again he will be surprised to find out that the number is divisible by 11. Ask him to pass the result to a third friend without you seeing it. Ask the third friend to divide this result by 13. Again, the number is divisible by 13. Take the result of this division without seeing it and pass it to the first person while telling him:

- Is this your number?

This beautiful arithmetic trick produces a magic impression on uninitiated audience. It is in fact very simple: Remember that attributed appending the initial three-digit number with the same three digits is equivalent to multiplying it by 1001, i.e. multiplying it by 7 x 11 x 13.

Chapter VI

The six-digit number which is received by your friend after appending it is therefore divisible by 7, 11, and 13, and then by successively dividing it by these three numbers (i.e. their product – 1001), we must obtain the original number again.

Shouldn't we marvel at this result as if we were a child reading the Scheherazade and its magical wonders of the Arabian Nights? The only difference is that any arithmetic miracle has a natural explanation, while the wonders of the East remain incomprehensible. Additionally, while arithmetic miracles are real, fairy tales miracles are fictitious...

Number 10101

After what has been said about the number 1001, you will not be surprised to see another window with number 10101:

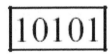

You can guess which proprieties have allowed this number to gain such honors. Like number 1001, it gives surprising results when multiplied by two-digit numbers (instead of three-digit numbers): each two-digit number, when multiplied by 10,101, resulting in itself, written three times. For example: $73 \times 10101 = 737373$, $21 \times 10101 = 212121$. The reason becomes clear from the following line:

$$73 \times 10101 = 73\,(10000 + 100 + 1) = \left\{ \begin{array}{r} 730000 \\ 7300 \\ + \quad\quad 73 \\ \hline 737373 \end{array} \right. .$$

Can I pull off some guessing tricks with the help of this extraordinary number like what I did using the number 1001? Of course, we can obtain even more spectacular impression if we bear in mind that 10101 is the product of four primes:

$$10101 = 3 \times 7 \times 13 \times 37.$$

First, you offer a person to think about a two-digit number (any of them) and write it down without you seeing it, you offer a second person to write down same number again, and then you offer a third person to write it down another time so that they obtain a six-digit number. You offer a fourth person to divide the resulting six-digit number by 7, then the fifth person should divide the resulting quotient by 3, the sixth person divides the resulting quotient (from the division by 3) by 37, and finally, the seventh person divides the resulting quotient by 13. All the four divisions are performed without them seeing the original number. Then you take the result of the last division and ask the first person whether this is the number he initially wrote down. He will be surprised.

You can repeat the same trick while adding some variety to it by using different dividers. Namely, instead of dividing by 3, 7, 13, 37, you can divide by 21, 13, 37 or 7, 39, 37 or 3, 91, 37, or 13, 7, 111.

This number – 10101 – probably possesses more magical proprieties than the number of Scheherazade (1001), even if it is less known than this one. Leonty Mignitsky has written about it about two hundred years ago in his "Arithmetic" book, and included some surprising calculations. It is one more reason for including it in our collection of curiosities arithmetic.

Six ones

The following figure illustrates another number that have some unusual properties that allow it to enter our museum.

$$\boxed{111111}$$

The number consists of six units. Thanks to our familiarity with the magical properties of number 1001, we immediately realize that:

$$111111 = 111 \times 1001.$$

However, 111 = 37 x 3 and 1001 = 7 × 11 × 13. It follows that our number (111111) is the product of five prime factors. Combining these five factors in two groups in various ways, we get 15 pairs of factors that give the same product (number 111111), namely:

$$3 \times (7 \times 11 \times 13 \times 37) = 3 \times 37037 = 111111$$
$$7 \times (3 \times 11 \times 13 \times 37) = 7 \times 15873 = 111111$$
$$11 \times (3 \times 7 \times 13 \times 37) = 11 \times 10101 = 111111$$
$$13 \times (3 \times 7 \times 11 \times 37) = 13 \times 8547 = 111111$$
$$37 \times (3 \times 7 \times 11 \times 13) = 37 \times 3003 = 111111$$

Yakov Perelman

$$(3 \times 7) \times (11 \times 13 \times 37) = 21 \times 5291 = 111111$$
$$(3 \times 11) \times (7 \times 13 \times 37) = 33 \times 3367 = 111111$$
$$\text{etc.}$$

This means that you can have 15 people each completing single but different multiplications of two pair of numbers and still obtaining the same result: 111111. Additionally, you can use this number for guessing games in the same manner as what we have done with numbers 1001 and 10101. In this case, you need to have someone think about a one digit number, and then ask him to repeat the digit 6 times. Have other people successively divide the result by the following five primes: 3, 7, 11, 13, and 37 (or other combinations such as 21, 33, 39, etc), you will obtain the original number. This allows various entertaining tricks.

Numeral pyramids

The following museum window shows numeric attractions of a special kind. They resemble pyramids composed of numbers. Here is a closer look at the first of these pyramids:

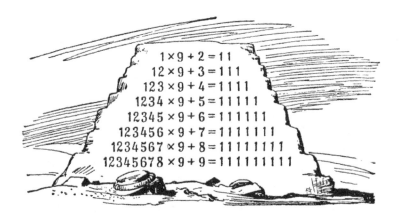

$$1 \times 9 + 2 = 11$$
$$12 \times 9 + 3 = 111$$
$$123 \times 9 + 4 = 1111$$
$$1234 \times 9 + 5 = 11111$$
$$12345 \times 9 + 6 = 111111$$
$$123456 \times 9 + 7 = 1111111$$
$$1234567 \times 9 + 8 = 11111111$$
$$12345678 \times 9 + 9 = 111111111$$

How to explain the peculiar results of these arithmetic operations and the resulting strange pattern?

Take for example one of the middle ranks of our numeric pyramid: 123456 × 9 + 7. Instead of multiplying by 9, we can equivalently multiply by (10 - 1), i.e. to add a 0 digit to the original number and subtract the same original number from the result. Thus, we have the following equality:

$$123456 \times 9 + 7 = 1234560 + 7 - 123456 = \begin{cases} \begin{array}{r} 1234567 \\ -\ 123456 \\ \hline 1111111 \end{array} \end{cases}.$$

Just look at the last subtraction to understand why we get the result above, which consists of one-digits only.

We can also understand this feature using other considerations. To transform 12345... into 11111..., we need to subtract 1 from the second digit, 2 from the third, 3 from the fourth, 4 from the fifth and so on. In other words, we subtract from the number 12345... the same number but ten

Yakov Perelman

times smaller with the last digit suppressed from it. Now it is clear that to get the desired result, we need to multiply the number by 10 and add to it the following last digit and subtract from it the original number (multiplying by 10 and then subtracting the multiplicand is equivalent to multiplying by 9).

The following numeric pyramid can be explained in the same manner:

$$1 \times 8 + 1 = 9$$
$$12 \times 8 + 2 = 98$$
$$123 \times 8 + 3 = 987$$
$$1234 \times 8 + 4 = 9876$$
$$12345 \times 8 + 5 = 98765$$
$$123456 \times 8 + 6 = 987654$$
$$1234567 \times 8 + 7 = 9876543$$
$$12345678 \times 8 + 8 = 98765432$$
$$123456789 \times 8 + 9 = 987654321$$

This one is obtained by multiplying numbers built using successive digits by 8 and then adding successively increasing numbers. The last line of this pyramid is particularly interesting. Indeed, by multiplying the number by 8 and adding 9, the original order of the digits in the number is completely reversed.

Obtaining such strange results becomes clear from the following line:[3]

---

3 The reason for 12345 × 9 + 6 giving 111111 is discussed in the previous numeral pyramid.

$$12345 \times 8 + 5 = \left\{ - \begin{array}{l} 12345 \times 9 + 6 \\ 12345 \times 1 + 1 \end{array} \right\} = \left\{ - \begin{array}{l} 111111^* \\ 12346 \end{array} \right. ,$$

i.e. 12345 × 8 + 5 = 111111 - 12346. But subtracting the number 12346 from 111111 leads to a number composed of a series of increasing digits (98765).

The validity of the third numeral pyramid reproduced here, is a direct consequence of the validity of the first two.

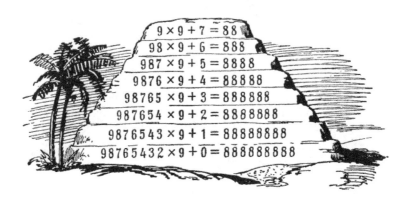

This connection is established very easily. For example, from the first pyramid, we already know that:

$$12345 \times 9 + 6 = 111111.$$

Multiplying both sides by 8, we obtain: (12345 × 9 × 8) + (8 × 6) = 888888.

But from the second pyramid, we know that 12345 × 8 + 5 = 98765 or 12345 × 8 = 98760.

Yakov Perelman

This means: $888888 = (12345 \times 9 \times 8) + (8 \times 6) = (98760 \times 9)$ $+ 48 = (98760 \times 9) + (9 \times 5) + 3 = (98760 + 5) \times 9 + 3 = 98765$ $\times 9 + 3.$

Now, you are sure the original numeral pyramid is not as mysterious as it seems at first glance. The calculations of their formation are not difficult to understand if you observe them closely. This did not prevent a German newspaper a few years ago from putting them in their columns with the following note: "The reason for these striking regularities is still not been explained." You can see that here, and affirm that there is almost nothing to explain.

Nine identical digits

From the last line of the first numeral pyramid above, we know that:

$$12345678 \times 9 + 9 = 111111111$$

This line explains a whole group of interesting arithmetic curiosities that have been collected in our museum in the following table:

```
12345679 ×  9 = 111111111
12345679 × 18 = 222222222
12345679 × 27 = 333333333
12345679 × 36 = 444444444
12345679 × 45 = 555555555
12345679 × 54 = 666666666
12345679 × 63 = 777777777
12345679 × 72 = 888888888
12345679 × 81 = 999999999
```

Taking into account the fact that $12345678 \times 9 + 9 = (12345678 + 1) \times 9 \times 9 = 12345679$. We obtain $12345679 \times 9 = 111111111$.

Hence it follows directly that

$12345679 \times 9 \times 2 = 222222222$
$12345679 \times 9 \times 3 = 333333333$
$12345679 \times 9 \times 4 = 444444444$
etc.

Digital staircase

What happens if the number 111111111, which we have seen before is multiplied by itself? You can predict in advance that the result should be outlandish – but do you exactly know it? If you have the ability to imagine figures of rows, you can find an interesting way to calculate the result without resorting to a multiplication on paper. After all, in essence, the multiplication here is limited to the proper alignment of digits, because all what is needed is to multiply one unit by unit. Then adding the results of different multiplications is reduced to the simple addition

Yakov Perelman

of units.[4] Taking into account the staggering of the nine rows of units, we can easily find - even without writing the table reproduced here below - the result of this one-of-a-kind multiplication: 12345678987654321.

```
            1 1 1 1 1 1 1 1 1
            1 1 1 1 1 1 1 1 1
            ─────────────────
            1 1 1 1 1 1 1 1 1
          1 1 1 1 1 1 1 1 1
        1 1 1 1 1 1 1 1 1
      1 1 1 1 1 1 1 1 1
    1 1 1 1 1 1 1 1 1
  1 1 1 1 1 1 1 1 1
1 1 1 1 1 1 1 1 1
```

1 2 3 4 5 6 7 8 9 8 7 6 5 4 3 2 1

All the nine digits are arranged in strict order, decreasing symmetrically from the middle in both directions. Readers who are tired of numeric curiosities can leave here this gallery and go to the next gallery of the arithmetic museum where giant numbers are exhibited.

Magic Ring

What a strange ring on display in the next gallery of our

---

4 In the binary system, as we have explained previously (see Chapter V), all the multiplications are of this type. In this example, we are clearly convinced of the advantages of the binary system.

museum? In front of us, we have three flat rings rotating one inside the other. On each ring, 6 digits are written in the same order. In other words, each ring contains the same number: 142857

$$\boxed{142857}$$

The reason that led us to put the rings in our arithmetic cabinet of curiosities is this amazing property they have: No matter how you rotate the rings, adding the two numbers written on the outer and the middle rings (the numbers are formed by starting from the same point and moving in the direction shown by the arrows in the figure), will lead to a six-digit number that could be formed on the inner ring (the result is generally a six-digit number). Try it for yourself, move the Magic Ring!

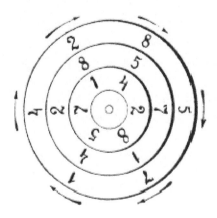

In the example shown in the above figure, adding the numbers in the two outer rings:

$$+ \quad \frac{\begin{array}{r} 142857 \\ 428571 \end{array}}{571428} ,$$

We obtain the number displayed on the inner ring. In another arrangement of the rings relative to each other, we have the following two examples:

$$+ \quad \frac{\begin{array}{r} 285714 \\ 571428 \end{array}}{857142} , \qquad + \quad \frac{\begin{array}{r} 714285 \\ 142857 \end{array}}{857142} .$$

There is, however, an exception to this general rule. Indeed, when the six digits in the outer ring complement the six digits in the middle ring, we obtain a sum equal to 99999 as illustrated in the following example:

$$+ \quad \frac{\begin{array}{r} 285714 \\ 714285 \end{array}}{999999} .$$

There is also another impressive property to the ring. If you subtract the number the outer number from the middle one, you will be able to obtain an inner one. Here are some examples:

$$- \quad \frac{\begin{array}{r} 428571 \\ 142857 \end{array}}{285714} , \qquad - \quad \frac{\begin{array}{r} 571428 \\ 285714 \end{array}}{285714} , \qquad - \quad \frac{\begin{array}{r} 714285 \\ 142857 \end{array}}{571428} .$$

Chapter VI

There is an exception to this rule. This one occurs when the two numbers (the outer and the middle ones) are perfectly aligned and their digits match each other. In this case the difference between the two is zero. But that are not all the wonderful qualities of our number (142857). Multiply it by 2, 3, 4, 5 or 6 and you will get, as before, the same original digits, with one or more of them moved in a circular manner:

$$142857 \times 2 = 285714$$
$$142857 \times 3 = 428571$$
$$142857 \times 4 = 571428$$
$$142857 \times 5 = 714285$$
$$142857 \times 6 = 857\,142$$

You can see that the product of the multiplication contains the same digits as the original number but the order of these ones is changed: A group of digits from one side is moved to the other side of the number.

It is time to explain the reason behind all the features of this mysterious number. We will start to unravel its secrets by multiplying it by 7: the result is 999999. So our number is nothing more and nothing less than one seventh of 999999, i.e., the fraction 142857/999999 is equal to 1/7. And if you convert 1/7 to a decimal fraction, you get:

$$1:7 = 0,142857 \ldots \quad \text{that is} \quad 1/_7 = 0,(142857).$$
$$\underline{10}$$
$$\quad \underline{30}$$
$$\quad\quad \underline{20}$$
$$\quad\quad\quad \underline{60}$$
$$\quad\quad\quad\quad \underline{40}$$
$$\quad\quad\quad\quad\quad \underline{50}$$
$$\quad\quad\quad\quad\quad\quad 1 \quad \cdot$$

Our mysterious number is, hence, an infinite periodic fraction, which is obtained by repeating 7 digits. It becomes clear now why we obtain such a feature when doubling it, etc. This number simply permutes a group of digits from one place to another place. When multiplying it by 2, we obtain the fraction 2/7 instead of 1/7. When you try to convert the fraction 2/7 into decimals, you will notice that2 is one of the reminders that we have obtained in the conversion of 1/7: it is clear that the digits obtained in the conversion of 2/7 would a repeat of the digits obtained in the conversion of 1/7. In other words, we should get the same period, but only a few digits will be different on the end. The same should happen when multiplying by 3, 4, 5, and 6, i.e., the same period would be found in the decimal conversion. When multiplying by 7, we get one whole unit – i.e., 0.999999.... if you think about 1 as an infinite periodic decimal.

The curious results of addition and subtraction on the rings are explained by the same fact (that a decimal fraction with a period 142857 is equal to 1/7). What we are doing by turning a ring a few numbers? We swap the front group of digits at the end, i.e., and therefore, as per the explanation above, we are multiplying the number 142857 by 2, 3, 4, etc. Consequently, all the operations of addition or subtraction of numbers written on the rings are reduced to the addition or subtraction of fractions 1/7, 2/7, 3/7, etc. As a result, we need to get, of course, a few seventh parts – i.e., once again our series of numbers 142857 in varying circular permutations. Hence it is necessary to exclude only the cases where when such numbers are added, we obtain seventh parts that add up to 1 or greater than 1.

However, these cases should not be completely excluded: they give results that are not identical to the previously discussed ones, although they are still very similar to them. Consider carefully what result should turn from the multiplication of our mysterious number by a factor of more than 7, i.e., 8, 9, etc. Multiply 142857, for example by 8: we can multiply first 7 (we will obtain 999999) and then add original number:

$$142857 \times 8 = 7 \times 142857 + 142857 = 999999 + 142857 = 1000000 - 1 + 142857 = 1000000 + (142857 - 1) = 1142856$$

The final result - 1142856 – differs from the original number -142857- by having a one unit digit appended to it on the left side addition to being decreased by one unit. We can use this similar rule to multiply 142857 by any other number that is greater than 7, as it can be easily seen in the following lines:

$$142857 \times 8 = 142807 \ (142857 \times 7) + 142857 = 1000000 - 1 + 142857 = 1142856$$
$$142857 \times 9 = 142857 \ (142857 \times 7) + (142,857 \times 2) = 1000000 - 1 + 285714 = 1285713$$
$$142857 \times 10 = (142857 \times 7) + (142857 \times 3) = 1000000 - 1 + 428571$$

Yakov Perelman

= 1428570

$142857 \times 16 = (142857 \times 7 \times 2) + (142857 \times 2) = 2000000 - 2 + 285714 = 2285713$

$142857 \times 39 = (142857 \times 7 \times 5) + (142857 \times 4) = 5000000 - 5 + 571428 = 5571427$

The general rule here is as follow: multiplying 142857 by a factor is equivalent to multiplying it by the decomposition of this factor into a multiple of 7 and a reminder the factor division by 7. When multiplying 142857 by a multiple of 7 is it possible to write the result in the form of subtraction between a multiple of million and the same multiple of unit.[5] Suppose that we want to multiply 142857 by 86. Decompose 86 by dividing it by 7. You will obtain 86 = 7 x 12 + 2. Therefore, the multiplication leads to 12000000 − 12 + 285714 = 12285702.

Another example: Let's consider the multiplication of 142857 by 365; we obtain (as 365 modulo 7 gives 52, and a remainder of 1)

$$142857 \times 365 = 52142857\text{-}52 = 52142803.$$

Having learned this simple rule and memorizing the multiplication results of our outlandish number by factors of 2 to 6 (which is not difficult - you need to remember only with which digits they start), you can astonish uninitiated people with the lightning speed of the multiplication of this six-digit number. In order to remember this number, you can use the following facts: The number comes from the division of 1 by 7. With this you can easily find the first three digits 142. The other three are obtained by

5 If the multiplier is a multiple of 7, the result is equal to the number 999999 multiplied by the number of sevens in the factor. This allows to easily perform the multiplication in mind. For example, $142857 \times 28 = 999999$ x 4 = 1000000 x 4 - 4 = 3999996.

subtracting the first three from three nines:

$$
\begin{array}{r}
999 \\
\underline{142857} \\
857
\end{array}
$$

We have already dealt with these numbers - namely, when we discussed the properties of number 999. Remembering what we have said in that section, we immediately realize that the number 142857 is obviously the result of multiplying 143 by 999:

$$142857 = 143 \times 999.$$

But $143 = 13 \times 11$. Remember what we have mentioned earlier regarding 1001, which is equal to $7 \times 11 \times 13$. We will be able to predict without performing any operations what should be the result of the multiplication 142857×7:

$$142857 \times 7 = 143 \times 999 \times 7 = 999 \times 11 \times 13 \times 7 = 999 \times 1001 = 999999$$

(We can certainly perform all these transformations in mind).

Phenomenal Family

We have seen that the number 142857 is a member of a whole family of numbers having the same properties. Here's another number: 0588235294117647

058823594117647

The 0 on the left is also part of the number and will play a role later. If you multiply this number by 4, you will get the same number but with the first four digits moved to the other side:

0588235294117647 x 4 = 2352941176470588

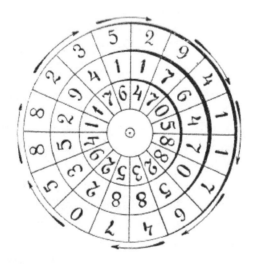

When represented on a wheel (as shown above) with three rings, this number exhibits the following property: When adding the numbers on the outer rings, we would be able to obtain the number in the inner ring, which is the same one but shifted in a circular manner. Here is an example:

$$+ \quad \begin{array}{r} 0588235294117647 \\ 2352941176470588 \\ \hline 2941176470588235 \end{array} .$$

This property is valid even when the two numbers to be summed are identical. Similarly, if we subtract the numbers on the first two rings, we would be able to get the number on the inner ring.

$$- \quad \begin{array}{r} 2352941176470588 \\ 0588235294117647 \\ \hline 1764705882352941 \end{array} .$$

Finally, this number consist of two halves: the second half is the difference between an 8 digit numbers containing nines and the first half. It is not hard to guess how the presented numeric series turned out a so close relative of the number 142857, if the last number represents the period of the infinite fraction $1/7$, then our number is likely to be a period of some other fraction. And it really is: our long series of digits is neither more nor less than the period of an infinite fraction, namely $1/17$:

$$1/17 = 0, (0588235294117647).$$

This is why the multiplication this number by multipliers from 1 to 16 1 turns out the same series of digits, in which only one or a few initial digits are transferred to the end of the number. And vice versa, transferring one or a few digits from the beginning of the number to its end is equivalent to increasing the number several times (1 to 16). Combining two rings that rotate with respect to one another, we make an addition of two multiples of this number, for example, the triple and the decuple of the number. Obviously, we should get the result on the ring because 13 times the original number is just another permutation of the original series of digits.

However, at a certain positions of the rings, we obtain numbers that are slightly different from the original series. For example, if we turn the ring so that the first number is 6 times the original number while the second number is 15 times the original, consequently the sum of the two numbers should be 6 + 15 = 21 times the original number. But such multiplier is easy to guess as we know the results of the multiplication by numbers between 1 and 16. Indeed, as our number period is equal to 1/17, it would give 16 nines (i.e., as much as their denominator impliedly recurring decimal) when multiplied by 17, or 1 followed by 17 zeros minus 1. Therefore, when our number is multiplied by 21, i.e., 17 + 4, with would get the multiplier associated with 4 added to 1 unit. The multiplier associated with 4 starts with the same digits that result from the conversion of the fraction 4/17 into a decimal fraction.

$$\begin{array}{l} \underline{4} \quad : \quad 17 = 0{,}23... \\ \underline{40} \\ \underline{60} \\ 9 \end{array}$$

Order of the other digits is known: 5294 ... Consequently, 21 times our number will 2352941176470588, which is the same number obtained from the sum of digits in circles according to their location. Obviously, when doing the

Chapter VI

subtraction instead of the addition, it is not possible to obtain this situation.

There are several numbers like those two. They all constitute a single family and they are united by a common origin which is the ability to transform them from a simple fraction into an infinite periodic decimal. But not every number having the decimal period discussed above gives a remarkable property of permutating the digits when multiplied. This is the case only for the fractions in which the number of digits in the period is smaller by one unit when compared to the denominator of the corresponding simple fraction. Examples include:

$1/7$ gives a period with 6 digits
$1/17$ «   «   «   16   «
$1/19$ «   «   «   18   «
$1/23$ «   «   «   22   «

You can do some testing to verify that the periods of fractions obtained from the conversion of 1/19 and 1/23 in decimal fractions, have the same features as we have discussed with the periods of the fractions 1/7 and 1/17. If the condition regarding the number of digits in the period is not met, then the corresponding period yields a number that does not belong to our family occupying of interesting numbers. For example, 1/13 gives a period six (rather than 12) digits:

$$1/13 = 0.076923.$$

Multiplying by 2, we get a completely different number:

Yakov Perelman

$$2/13 = 0.153846.$$

Why? Because of the reminders of the division 1:13 number was not one of them. The number of the reminder was as much as the number of digits in the period, i.e., 6.

We have 12 different multipliers for the same fraction 1/13, and consequently, not all the factors will be among the remainders (in fact only 6 factors will be in this case). It is easy to see that these factors are the following: 1, 3, 4, 9, 10, and 12. The associated multipliers give 6 cyclic permutation (076923 × 3 = 230 769), but not the others. That is why only a limited number turns in the "magic ring" give the desired result and not all the turns.

CHAPTER VII

Tricks Without Cheating

King Rituparna art

Arithmetic tricks - honest conscientious tricks - do not attempt to cheat or lull the audience attention. To perform an arithmetic trick, you don't need any miraculous dexterity, amazing physical agility, or artistic abilities that sometimes require perennial exercise. The whole secret of any arithmetic trick is to use the curious properties of numbers, in close acquaintance with their peculiarities. For these who know the answer to such tricks, all seems simple and straightforward, but for unaware people, a simple arithmetic operation such as multiplication seems to have a bit of magic in it.

There was a time when doing the most mundane arithmetic operations on large numbers, which are now well-known by every schoolboy, was an art that only a few people were able to master. They seemed to possess some sort of supernatural abilities.

In "Nala and Damayanti" story, we find an echo of this view of arithmetic. King Rituparna proved to his disciple Nala his ability to instantly count the number of leaves on a thick canopy covered tree, and agreed to show him the secret of this art.

The secret of this art can be explained as follow: Counting the leaves one by one would be extremely time-consuming. So, instead of counting the leaves, the king counted the branches and then multiplied that number by the number of leaves in a branch (assuming that equally grown branches contain the same number of leaves). The operation of multiplication was so unfamiliar to people at that time that it looked like something mysteriously supernatural.

King Rituparna and Nala

The answer to most of the arithmetic tricks is as simple as the secret of King Rituparna "trick". One has to only know the solution to a trick and he will be able to learn the art of using and presenting it. At the heart of each arithmetic trick lie some interesting features of numbers. Learning these features is very instructive in addition to being entertaining.

Sealed envelopes

A Magician pulls out a stack of 300 one-ruble banknotes, along 9 envelopes. He requests you to distribute the money among the nine envelopes, so that he can call any number between 1 and 300 and you should provide it to him using a subset of envelopes without moving notes from an envelope to another.

The task seems completely daunting to you. You already think that the magician will try to trick you using some artful wordplay or an unexpected interpretation of words' meanings. Seeing your helplessness, he lays out the money in the envelopes, seals them, and offers you to select any amount between 1 and 300 rubles.

Let's suppose you have randomly selected number 269.

Without the slightest delay, the magician selects 4 sealed envelopes. You open them and find:

In the 1-E	—	64	rubles
« 2-E	—	45	«
« 3-E	—	128	«
« 4-E	—	32	«
Total		269	rubles

Now, you tend to suspect a skilful magician who has the necessary experience substituting envelopes. He calmly puts the money back into the envelopes, seals them again. You call a new number such as 100, or 7, or even 293, and the magician instantly identifies the envelopes that need to be pulled in order to reach the required amount (for 100 rubles, he would pull 4 envelopes, for 7 rubles, he would pull 4 envelopes, and for 293 rubles, he would pull 6 envelopes).

The above may seem incomprehensible, but after reading this paragraph, you will be able to repeat the same trick

Chapter VII

and amaze your friends. The secret of this trick lies in the fact that the banknotes are divided between the envelopes as follow: 1 r., 2 r., 4 r., 8 r., 16 r., 32 r., 64 r., 128 r. and finally, in the last – the remaining rubles, i.e., 300 - (1 + 2 + 4 +8 + 16 + 32 + 64 + 128) = 300-255 = 45.

The first 8 envelopes allow easily to make any amount between 1 and 255, but if you are given a larger number, then the last envelope enters into action. When the 45 rubles are subtracted from the number, the result can be built using the first 8 envelopes.

You can check the suitability of such grouping using numerous examples and make sure all of them can be obtained using the 9 envelopes. But you are probably wondering why a series of numbers 1, 2, 4, 8, 16, 32, 64, etc. has such a remarkable property. This is easily understood if you remember that the numbers of our series represent powers of two: 2^1, 2^2, 2^3, 2^4, etc.,[1] and therefore, each number can be regarded as a level in the binary system. Since every number can be written in a binary system, this means then every number may be decomposed as the sum of powers of two, i.e., a series of numbers 1, 2, 4, 8, 16, etc. So when you pick up envelopes you are simply expressing the specified number in the binary system. For example, number 100 can be easily decomposed in the binary system as follow:

[1] We know in algebra that the number 1 can be viewed as a power of 2, namely 2^0.

Yakov Perelman

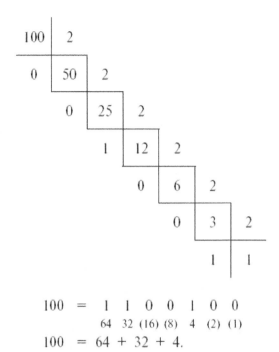

$$100 = \begin{array}{ccccccc} 1 & 1 & 0 & 0 & 1 & 0 & 0 \\ 64 & 32 & (16) & (8) & 4 & (2) & (1) \end{array}$$

$$100 = 64 + 32 + 4.$$

(Recall that in the binary system, the first place to the right contains units, the second place contains the twos, the third place the fours, etc)

Predicting the number of matches in a box

The same property of the binary system can be used for this next trick. Offer someone to loan you a full matchbox. Put it on the table, and next to it put one after the other 8 squares of paper organized in two rows. Then ask him in your absence to do the following: Leave half of the matches in the box, and move the other half to the nearest square of paper on the right, if the number of matches is odd, then put the excess match on the square of paper to the left of the first one.

Now, divide the matches that are on the first square of paper into two halves (without touching the match next to them on the left if any). Move the first half back to the box and move the other half to the next square. In case of an odd number, move the remaining match to the square of paper on the left. Repeat the same process again while not forgetting, in case of an odd number of matches, to put one match on the square of paper to the left. In the end, all the matches except the ones lying on the left squares are returned to the box.

When this process has been completed, and you are back in the room, have a glance at the blank piece of paper, and calculate the original number of matches in the box.

This trick usually amazes the uninitiated who could not understand how it is possible to guess the number of matches in the box using a blank piece of paper.

In fact, the "empty" paper is very eloquent in this case: Combined with the remaining matches, you can literally read on it the desired number. It can be seen as a number written in the binary system. Let's explain this with an example. Let suppose the number of matches in the box was 66. The sequential operations that are carried out on them lead to the paper that is shown in the following two figures:

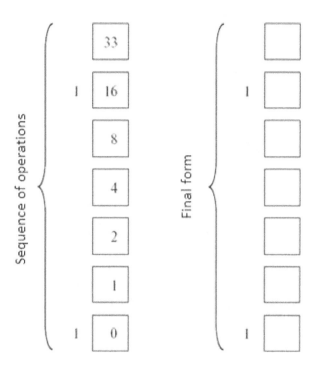

You don't need to think too much to realize that what has been accomplished is to express the original number of matches into the binary system. The above figure on the right depicts the resulting binary number. If there is a blank paper on the left, it means 0 in the binary system, it there is a mark indicating a match to the left, it means a 1. Reading from the bottom to the top of the figure, we obtain:

1	0	0	0	0	1	0
64	(32)	(16)	(8)	(4)	2	(1)

Chapter VII

That is 64 + 2 = 66 in the decimal system.

If the matchbox contained 57 matches, we would have obtained the following figure, i.e., number 57 written in the binary system:

$$\begin{array}{cccccc} 1 & 1 & 1 & 0 & 0 & 1 \\ 32 & 16 & 8 & (4) & (2) & 1 \end{array}$$

We can check the result in the decimal system as follow: 33 + 16 + 8 + 1 = 57.

As a variation, you can also use two or more matchboxes and guess the amount contained in them matches.

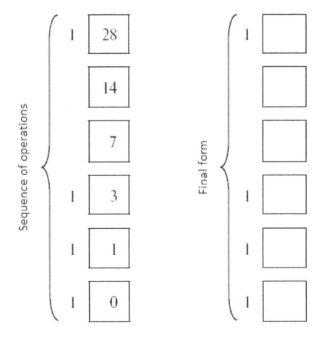

Reading thoughts on matches

The third variation of the same trick allows to find a number from a representation that use matches. First, you think of a number, and then you divide it by 2, if the resulting number is even, you place a horizontal match, and then divide the result again by 2, and start over. If the resulting number is odd, you place a vertical match, subtract 1, and then start over. By the end of all the operations, you obtain a figure like the following one:

You look into this shape and unmistakably identify number 137. How do you know it? The answer becomes clear if you consistently denote near each match the number obtained from the division:

$$
\begin{array}{c c}
68 & 34 \\
\hline
137 \;\Big| & 17 \;\Big|\underline{}\Big| \; 1 \\
& 8 \quad 4 \quad 2
\end{array}
$$

Since the last match represents number 1, it is not difficult, going from there to the preceding divisions, to get to the original number. For example, in the following figure you can identify number 664.

In fact, consistently doubling the number (starting from the right side) and not forgetting to add a unit in the appropriate times, we get:

$$
\begin{array}{rcl}
| & = & 1 \\
- & = & 2 \\
| & = & 5 \\
- & = & 10 \\
- & = & 20 \\
| & = & 41 \\
| & = & 83 \\
- & = & 166 \\
- & = & 332 \\
- & = & 664 \\
\end{array}
$$

Thus, using matches, you are tracing the course of the thoughts of others, rebuilding their entire chain of reasoning.

The same result can be obtained using another approach: realizing that a horizontal match corresponds to a binary zero (division by 2 without remainder), and a standing one corresponds to a binary one, you discover that in the above example, you are in fact reading (from right to left)

Yakov Perelman

the following binary representation of the number:

$$1 \quad 0 \quad 0 \quad 0 \quad 1 \quad 0 \quad 0 \quad 1$$
$$128 \, (64) \, (32) \, (16) \, 8 \, (4) \, (2) \, 1$$

which leads to the following number in the decimal system:

$$128 + 8 + 1 = 137.$$

In the second example the desired number is represented in the binary system as follow:

$$1 \quad 0 \quad 1 \quad 0 \quad 0 \quad 1 \quad 1 \quad 0 \quad 0 \quad 0$$
$$512 \, (256) \, 128 \, (64) \, (32) \, 16 \, 8 \, (4) \, (2) \, 1$$

Which can be written in the decimal system as follow: 512 + 128 + 16 + 8 + 1 = 664.

Another example: Which number can you identify from the following figure of matches?

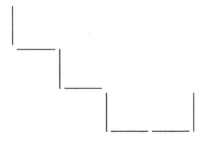

Answer: The figure illustrates 10010101 in the binary system. The decimal representation is:

Chapter VII

$$128 + 16 + 4 + 1 = 139.$$

It should be noted that because the first match is a standing one, the number is odd in the decimal system.

Ideal weights

Some readers may probably have the following question: Why have we considered the binary system in the above problems? After all, every number can be represented in any system, and among them, the decimal one. Why may the binary be preferred sometimes?

Well, in the binary system, if we put the zero aside, only one digit is used - a one, and consequently, for each digit in each position, we have only two possibilities (either one or zero). If in the trick with the envelopes, we distributed the money, for example, in the base 5 system, it could be possible to make any amount without opening the envelopes, only when every pack is repeated to us at least 4 times (the base 5 system uses only 4 digits, zero aside).

However, there are times where it is more convenient to use the binary or ternary systems in a somewhat modified version. This includes the famous old "problem of a system of weights", which can serve as the plot of an arithmetic problem.

Imagine that you are asked to come up with a system of four weights, with which it would be possible to weigh any whole number of pounds from 1 to 40. With a binary system you would come with: 1 lb., 2 lbs., 4 lbs., 8 lbs., 16 lbs. With these weights you measure any weight from 1 to 31 pounds. This obviously does not satisfy the required conditions: The number of weights is 5 and not 4, and the number of possible weights is 31 and not 40. However, on the other hand, you have not used all the solutions allowed by the weights here. It is not only possible to add up the weights, it is also possible du subtract them (when they are placed on the opposite pan). This will allow you so many different combinations that you are simply lost in your quest to fit them into a numeral system. You may even doubt that this problem has problem has a solution when considering a number of weights as low as four. But a dedicated search would allow you to find the following four magical weights:

1 lb., 3 lbs., 9 lbs., 27 lbs.

Any weight between 1 and 40 pounds with an integer number of pounds can be weighted with these weights if they are placed properly on the correct pan. We will not even bother to give examples, because for each given weight it would be very easy to find the solution. But let's better understand why these four weights have this wonderful property. A careful reader would have probably noticed that the numbers associated with these weights are series of powers of 3.[2]

$$3^0, 3^1, 3^2, 3^3.$$

In other words, we are using here the help of a ternary notation. But how to use it in cases where the desired weight is obtained as the difference between two weights? And how to avoid the need for the use of all possible values such as the doubles (in the ternary system, in addition to

2 The unit in this system can be considered as 3 to the power of 0 (in general the base of the system to the power of 0).

zero, there are only two digits: 1 and 2)? The answer to these two questions is obtained by the use of "negative" numbers. For example instead of using number 2, it is possible to use 3 - 1, i.e., a higher digit unit with lower digit subtracted from it. For example, the number 2 in our modified ternary system will not be designed by a 2, but instead will be designed as $1\bar{1}$ where the minus sign above the units' digit indicated that the associated number of units is subtracted instead of being added.

Similarly, number 5 representation in this modified ternary system is not 12 but $1\bar{1}\bar{1}$ (i.e. 9 - 3 - 1 = 5). The first ten numbers representations in this modified ternary system are as follow:

1	2	3	4	5	6	7	8	9	10
1	$1\bar{1}$	10	11	$1\bar{1}\bar{1}$	$1\bar{1}\bar{0}$	$1\bar{1}1$	$10\bar{1}$	100	101

It is now clear that if any number can be represented in this extended ternary system using zeros (i.e., the absence of a number) and units that are added or subtracted from it then the numbers 1, 3, 9, 27 can be used, through addition and subtraction, to obtain all the numbers between 1 and 40. The cases of additions are obtained by the putting the weights on the scale pan opposite to that of the load. The cases of subtraction are obtained by placing the weights in the same scale pan as the load, and therefore the associated amount is subtracted from the amount of the weight on the opposite pan. Zero corresponds to the absence of weight.

Is this system used in practice? As far as we know, no. Everywhere in the world, where the metric system has been introduced, we weight things using 1, 2, 2, 5 units

Yakov Perelman

instead of 1, 3, 9, 27 - although the first can only weigh loads up to 10 units, and the second - up to 40. Even in place where the metric system has not been introduced and is not in use, the set of 1. 3, 9, 27 is not applicable. The reason for failing to use these weights in practice lies in the fact that they are only good on paper, they very troublesome to use in reality.

If it was necessary to use a predetermined number of weights to weight a given load (for example, 400 g of oil or 2500 g of sugar) then a system of weights built with 100, 300, 900, 2700 g would have a practical use (although looking for the appropriate combinations may take a long time). But when you have to add different amounts to obtain the load weight, such a system is terribly inconvenient: It is often preferable, for the sake of simplifying the additions, to have additional weights that are simply the sums of the above units. Weighing under such conditions becomes extremely slow and tediously though. You can easily see this if you write the above weights on pieces of papers, and try to weight some sample loads.

Predicting an unwritten number

We are struck by the ability of some people to add multi-digit numbers with extraordinary rapidity. But what about a person who can write down the sum even before he had heard the terms? This trick is usually performed the following way:

The guesser offers you to write down a multi-digit number of your choice. Glancing at this first term, the solver writes down on paper a future sum and hides it from you in your pocket. He will ask you (or someone from the audience) to write another term - again, a number of your choice. And then he quickly writes down the third term. You add up all the three terms - and get exactly the result that was previously written by the solver on a piece of paper and hidden from you. For example, if you wrote down 83267 as a first number, the solver writes down the future number 183266. Then you write, say, 27935, and solver immediately adds the third term - 72064:

I .... YOU :	83267
III .... YOU :	27935
IV .... GUESSER :	72064
II. ...... SUM :	183266

The solver has been able to find immediately to the third term that led to the result, even if he did not know the second term beforehand. This trick also works with 5 or 7 terms, but in this case, the guesser will write two or three of the terms. Here, the substitution of the paper containing the final result (as you may suspect) is not possible, because this one is stored in your own pocket until the last moment. Obviously, the solver uses here some unknown property of numbers.

This property is explained here: The solver uses the fact that the addition of, say, a 5-digits number to a five nines (99999), the first number is increased by 1,000,000 − 1, i.e., a digit of one unit is added to its left side and one unit is subtracted from its right side. For example:

Yakov Perelman

$$
\begin{array}{r}
83267 \\
+\ 99999 \\
\hline
183266
\end{array}
$$

It is this result – i.e. the sum of the number you have written and 99,999 – that is written on the piece paper as it will be the final result of the addition of the three terms. Now, in order to reach this number, he simply needs to add to your second term the necessary amount (in the form of the third term) in order to get 99,999, i.e. subtract from 9 each of your second term digits and write down the result. Applying this technique, you can easily find the third term in the above example as well as the following examples:

I . . . . YOU:	379264	I . . . . YOU:	9935
III . . . . YOU:	4873	III . . . . YOU:	5669
IV . . . . . GUESSER:	995126	IV . . . . . GUESSER:	4330
II . . . . .SUM	1379263	II . . . . .SUM	19934

You can easily notice that you can severely hamper the solver if the second term contains more digits than the first one. In this case the solver cannot write a third term that when added to the second one would lead to a smaller number (i.e. 99,999). In order to avoid this situation the solver usually restricts your freedom of choice, so that the second number contains less or equal digits to the first one.

This trick becomes impressive when it involves several persons. Let consider the case of 5 terms. After the first term - for example, 437692, the solver already predicts the sum of all the five numbers, namely 2,437,690 and records it (here he will add twice 999999, i.e. 2000000 - 2). The resolution is made clear in the following figure:

I.............	YOU WROTE :	437692
III..... THE SECOND PERSON WROTE :		822531
V....... THE THIRD PERSON WROTE :		263009
IV ....GUESSER ADDED :		177468
VI............. «	«	736990
II............. «	PREDICTED :	2437690

Predicting the outcome of a number of operations

Huge impressions can be made using arithmetic tricks in which the solver guesses the result of a set of operations even if he does not know the original numbers. There are many such tricks, and they are all based on the ability to come up with a number of arithmetic operations, from which the outcome does not depend on the original number.

Here is an example of such tricks:

The criterion for divisibility by 9 is well known to all: A number is a multiple of 9 if the sum of its digits is divisible by 9. Remembering this rule, we can formulate the following interesting proposition: The difference between a number and the sum of its digits is always a multiple of 9. For example: 457 - (4 + 5 + 7) = 441 is a multiple of 9. Similarly, the difference between two numbers containing the same digits is divisible by 9. For example 7843-4738 = 3105, a multiple of 9.

Yakov Perelman

You can use the proposition above to build a very simple trick. Invite a friend to think about a number, any number, then ask him to rearrange the digits in any order he wants, then ask him to subtract the smaller number from the larger one. In the resulting difference, ask him to hide one of the digits, any digit,- and to tell you loudly the rest of the digits. Since you know that the result is divisible by 9, you can quickly mentally figure out the missing digit as you can simply sum up the digits read loudly by your friend. For example: your friend think of number 57924: 92457 obtained after rearrangement.

$$\begin{array}{r} 92457 \\ - \underline{57924} \\ 3?533 \end{array}$$

Your friend crossed out a digit from the difference (indicated by a question mark). By summing the remaining digits 3 + 5 + 3 + 3, we get 14. It is not hard to figure out that the digit that was crossed out was 4 because the next highest multiple of 9 is 18 and 18 - 4 = 14.

This same trick can be done much more effectively, with the guessing of the entire number, and without doing any mental calculations. To perform this, the easiest way is to propose to your friend to think about a three digit number with unequal digits, then, to rearrange the numbers in reverse order, subtract the smaller number from the larger. In the resulting number, let him rearrange the numbers again and add both numbers. The final result of all these operations (a permutation, a subtraction, a permutation, and then an addition) will be known to you without the slightest delay. You can even give him in advance the result in a sealed envelope.

The trick's secret is very simple: no matter the number set initially, these actions always lead to the same number: 1089. Here are some examples:

$-\ \dfrac{\begin{array}{r}762\\267\end{array}}{}$	$-\ \dfrac{\begin{array}{r}431\\134\end{array}}{}$	$-\ \dfrac{\begin{array}{r}982\\289\end{array}}{}$	$-\ \dfrac{\begin{array}{r}291\\192\end{array}}{}$

$$
\begin{array}{r} 762 \\ -\ 267 \\ \hline +\ 495 \\ 594 \\ \hline 1089 \end{array}
\qquad
\begin{array}{r} 431 \\ -\ 134 \\ \hline +\ 297 \\ 792 \\ \hline 1089 \end{array}
\qquad
\begin{array}{r} 982 \\ -\ 289 \\ \hline +\ 693 \\ 396 \\ \hline 1089 \end{array}
\qquad
\begin{array}{r} 291 \\ -\ 192 \\ \hline +\ 099 \\ 990 \\ \hline 1089 \end{array}
$$

(The last example shows how your friend should act when the resulting difference is a two digit number instead of a three digit one.)

Observing more closely the process of the above calculations, you, no doubt, understand the reason behind this uniformity of results. When subtracting, you inevitably must obtain a digit 9 in the position of the dozens and around this digit, two digits (on either side) whose some is 9 as well. In the subsequent addition, it is clear that digit 9 must be obtained in the right side. Furthermore, in the dozens position, we will have digit 8 as the result of adding 9 + 9. Keeping the resulting extra hundred in mind combined with a 9 that results from the addition of the digits in the hundreds position, we obtain a 10. Hence - 1089.

If you will repeat this experiment several times in a row without making any changes to it, then your secret will, of course, be discovered: Your audience will think that the same number 1089 constantly turns out, though, perhaps, they will not realize in the reason of such constancy. So you need to alter the trick. Make it easy, since 1089 = 33 x 33 = 11 × 11 × 3 × 3 = 9 x 121 = 99 × 11. You need only, when you reach number 1089, to divide the result by 33, or 11, or 121, or 99, or 9 - and then only to call the audience to get by. You, therefore, has left 5 trick changes - not to mention the fact that you can ask the audience also to multiply the resulting amount, mentally performing the same action.

Instant division

There are several varieties of this type of tricks. We will describe one here. It is based on the following familiar property: When a multiply a number by a number consisting of a series of nines with both having the same number of digits, we obtain a result consisting of two halves: the first half represents the original number minus one, the second one represent the result of the subtraction of the first half from the nine series number. For example: $247 \times 999 = 246753$, $1372 \times 9999 = 13718628$, etc.

The reason for this property is easy to see from the following line:

$$247 \times 999 = 247 \times (1000 - 1) = 247\,000 - 247 = 246\,999 - 246.$$

Using this property, you can offer to a group of people a race where you and they have to calculate the results of a set of divisions: (a) 68933106: 6894, (b) 8765112348: 9999, (c) 543456: 544, (d) 12948705:1295, etc. Before they even could start the race, you would be able to present them the results: (a) 9999, (b) 87652, (c) 999, (d) 9999, etc.

Chapter VII

My favorite digit

Ask someone to name his favorite digit. Let's say he selected 6.

- That's amazing! - You exclaim. - This is just the most wonderful of all the digits.
- Why is it? - Your puzzled interlocutor would inquire.
- Here's an example: multiply your favorite digit (6) by 9 and then multiply the resulting number (54) by the factor 12345679:

$$12345679$$
$$\times \quad 54$$

What would be the result of this multiplication? Your friend performs multiplication - and gets an amazing result, consisting entirely of his favorite digit:

$$6666666666$$

- You see you have managed to multiply the number of your favorite digit. What a wonderful property!

But you will have the same property if your friend had selected any other of the nine digits, because each of them has this same property:

12345679	12345679	12345679
$\times\ 4 \times 9$	$\times\ 7 \times 9$	$\times\ 9 \times 9$
4444444444	777777777	999999999

Yakov Perelman

In order to understand the secret of this property, you just need to recall what we have said about number 12345679 in the "Museum of numeric curiosities."

Guessing a birthday

Here is an example of guess tricks that can be used in different ways. The description of the mechanism behind the trick is rather complex, however, this one produces spectacular results.

Suppose that you were born on the 18th of May, 1903, and that you are 20 years old now. Suppose that I don't know your birth date or your age. Nevertheless, I dare to guess both, by forcing you to do only a certain number of calculations. Namely: I will ask you to multiply the month's number (May, the 5th month) by 100, then to add to the product, the day in the month (18), double the amount, add 8, multiply the result obtained by 5, add 4 to the product, multiply the result by 10, and add 4 to the resulting number, and finally add your age (20) to the result.

When you are done with all this, tell me the final result of the calculations. I subtract it from 444. The resulting number allows me to guess your birthday as follow: From right to left, the obtained successive 2 digits represent respectively your age, the day, and the month of your birth. Let's perform these calculations:

$$5 \times 100 = 500$$
$$500 + 18 = 518$$
$$518 \times 2 = 1036$$
$$1036 + 8 = 1044$$

$$1044 \times 5 = 5220$$
$$5220 + 4 = 5224$$
$$5224 \times 10 = 52240$$
$$52240 + 4 = 52244$$
$$52244 + 20 = 52264$$

Performing the subtraction 52264 - 444, we obtain the number 51820. Now let's divide this number from right to left taking two digits at the time. We have: 5-18-20, i.e., your age (20 years), your day of birth (18), and your month of birth (the 5th month, May).

The secret of this trick can be easily understood by considering the following equation:

$$\left\{ \left[(100m + t) \times 2 + 8 \right] \times 5 + 4 \right\} \times 10 + 4 + n - 444 = 10000m + 100t + n.$$

Here the letter m denotes the month, t - the number of days, n - age. The left-hand side of the equation expresses all the operations you have consistently made, and the right-hand side expresses what happens if you open the brackets and you make all the simplifications.

In the terms of 10000m + 100t + n, the numbers m, t, and n cannot be more than two-digit numbers, so the number obtained as a result, should always, when its face is divided in three parts containing two digits each, split in the required numbers m, t, and n. If he wishes, the reader can come up with other inventive modifications of this trick, i.e., other combinations of operations that would lead to similar results.

Yakov Perelman

One of Magnitsky's "consoling actions"

In this section, we will disclose the secret of another uncomplicated trick. This one is described in detail in Magnitsky's "Arithmetic", in a chapter called "On consoling actions through the use of arithmetic operations."

This trick lets someone guesses a number in relation with money, days, hours, or "whatever that can be counted." Let's consider the example of a ring worn at the 2nd joint of the small finger (i.e. the 5th finger, or pinky) by the 4th person in a group of 8 people. The guesser has to find which of the eight people (numbered from 1 to 8) has the ring, and on which finger and which joint is this one worn?

"He will say: Please take the number of the person who has the ring. Double that number. Then to the resulting number add 5. Then multiply the result by 5 and add the result the number of the finger where the ring is worn. Multiply the result by 10, and add the number of joint where the ring is worn, and then tell me the resulting number.

The group of people met without the guesser, executed the instructions and calculated the resulting number. The obtained 702, and communicated this number to him. He subtracted 250 from it and obtained the result 452, i.e. the 4th person, the 5th finger 2nd joint."

$$
\begin{array}{rl}
4 & \text{persons} \\
\underline{2} & \text{multiplication} \\
8 & \\
\underline{5} & \text{addition} \\
13 & \\
\underline{5} & \text{multiplication} \\
65 & \\
\underline{5} & \text{addition of finger} \\
70 & \\
\underline{10} & \text{multiplication} \\
700 & \\
2 & \text{addition of joint} \\
702 & \\
\underline{250} & \text{subtraction} \\
452 &
\end{array}
$$

Do not be surprised that this arithmetic trick was known 200 years ago: The problem is quite similar to the ones already included in the first collections of mathematical entertainments compiled by French mathematician Bachet de Meziriac in his 1612 book "Entertaining and enjoyable numeric problems". As a general rule, it should be noted that of the mathematical games, puzzles and tricks that are prevailing in our time are of very ancient origin.

CHAPTER VIII
Quick Calculations and Perpetual Calendar

You have probably heard or even attended mathematics sessions where "brilliant mathematicians" mentally calculate with extreme speeds the number of weeks, days, minutes an second that passed since your birth, or determine the day of the week in which you were born, or even the what day of the week will be a given day in the future, etc. To perform most of these calculations, it is not necessary, however, to have extraordinary mathematical ability. After a short exercise, we can all develop such ability. We need only to know the secrets of these tricks. We will now present these ones.

"How many weeks is my age?"

To quickly determine the number of weeks in a given number of years, you only need to multiply the number of years by 52, i.e., on the number of weeks in a year.

Let suppose the number of years is 36. Multiplying 36 by 52, you can immediately tell the result (1872) without heavy calculations. How did we do that? Quite simply: 52 consists of 50 plus 2, 36 multiplied by 5, through bisection, gives 180, thus we know the two first left digits of the result are 18; further multiplication of 36 by 2 gives 72; thus the total final result: 1872.

It is easy to understand the result. Multiplying by 52 is equivalent to multiplying by 50 and 2, but, instead, of multiplying by 50, we multiplied by 100 and took the half – but because the original number (36) is even, the result will still have two zeros to the right. Thus we need only to replace these two zeros by the two digits resulting from the multiplication of 36 by 2.

Chapter VIII

Now it is possible to understand why the calculation was "brilliant" and the answer was very quick. However, you should not forget that the number of days in a year is 365 days, i.e., 52 weeks and 1 day. Consequently, for every 7 years, you need to add an extra week to the result.[1]

"How many days is my age?"

If you don't know the number of days in a given number of years, the resort to the following solution: Half the number of years, multiply by 73, and add a zero to the right - the result is the desired number of days (this formula becomes clear if we note that $730 = 365 \times 2$). If I am 24 years old, the number of days is obtained by multiplying $12 \times 73 = 876$ and adding a zero to the right- 8760. The multiplication by 73 can also be done in an abbreviated manner, as it will be discussed later.

The above result needs to be amended by a few days due to leap years. In order to do so, you need to add to the result of a quarter of the number of years (in our example, $24:4 = 6$; the overall result is therefore 8766).

"How many seconds is my age?"

This question[2] can also be answered pretty fast, using the following method: half the number of years, then multiply the result by 63, then multiply the same half by 72, put the

---

1 It is easy to calculate the number leap years and correct the end result for these ones.

2 The reader can devise a technique for the calculation of the number of minutes in an age, using the same approaches described in this chapter.

Yakov Perelman

result of this second calculation next to the first, and credit the result with three zeros to the right. If, for example, you are 24 years, to determine the number of seconds, you can proceed as follow:

$63 \times 12 = 756$ and $72 \times 12 = 864$, hence the result: 756 864 000.

The method described above is simplified as much as possible and is consequently very fast. Consequently, the reader is advised to make the same calculation using the classic ordinary way, to understand the huge time savings obtained from the use the method above.

As with days and weeks calculations above, the leap years are not taken into account. This approximation is not significant as we are dealing with numbers equal to hundreds of millions.

Regarding the correctness of our formula, the verification is very simple. To determine the number of seconds that are in a given number of years, you need to find the number of second in a year, i.e., $365 \times 24 \times 60 \times 60 = 31536000$. We will proceed using the same manner. 31536 is a big factor that can be split in two parts (31500 and 36). So instead of multiplying the number of years (24 years) by 31536, we can multiply the number by 31500 and multiply it by 36. But in order to complete these two multiplications we can simplify them further for convenience as illustrated in the following formulas:

$$24 \times 31536 = \begin{cases} 24 \times 31500 = 12 \times 63000 = 756000 \\ 24 \times \quad 36 = 12 \times \quad 72 = \quad 864 \end{cases}$$
$$756864$$

Thus, it remains to write down three zeros to the right in order to obtain the number of seconds.

Rapid multiplication techniques

We mentioned earlier that in order to perform complex multiplications, it is possible to divide them into multiple individual but easier actions. But there are also some other techniques that can be used to achieve the same objective and it is advisable to remember them so they can be used in resolving conventional calculations. For example a technique called cross-multiplication is very convenient when considering the multiplication of two-digit numbers. This method goes back to the Greeks and Hindus in ancient times. They called it the "lightning method" or "cross multiplication".

Let suppose we want to complete the following multiplication: 24 × 32. We can mentally arrange the digits as follow, one below the other:

$$\begin{matrix} 2 & & 4 \\ | & \times & | \\ 3 & & 2 \end{matrix}$$

Now, we complete the following operations:

(1) 4 × 2 = 8 - this is the last digit of the result.

(2) 2 × 2 = 4, 4 × 3 = 12, 4 + 12 = 16, 6 - the penultimate digit of the result, one to remember.

(3) 2 × 3 = 6, adding the remaining unit from the above operation, we have 7 - this is the first digit of the result.

All the digits are now known: 7, 6, 8 - 768.

After a brief surprise, this exercise is very easy to digest.

Another method, consisting in the use of so-called add-ons, is conveniently used in cases when the multiplied numbers are close to 100.

Let suppose that we want to complete the following multiplication: 92 × 96. The "supplement" for 92 to 100 will be 8 and the "supplement" of 96 will be 4. The actions required in the method are described as follow:

(1) The two first to the right digits are obtained by multiplying the two supplements.

(2) The next two digits are obtained by simply subtracting the supplement from the other term, i.e., subtracting 4 from 92 and 8 from 96. In either case we have 88.

(3) Consequently, the result is 8832.

That the result is correct can be clearly seen from the following transformations:

$$92 \times 96 = \begin{cases} 88 \times 96 = 88\,(100 - 4) = 88 \times 100 - 88 \times 4 \\ 4 \times 96 = 4\,(\ 88 + 8) = \ 4 \times \quad 8 + 88 \times 4 \end{cases}$$
$$\overline{\quad 92 \times 96 \qquad = \qquad 8832 \quad + 0}$$

There is also a method for rapid multiplication of three-digit numbers, it also saves a lot of time, but its use is more difficult and requires some mental effort as you have to keep in mind a few numbers at the same time.

What day of the week?

We will analyze now the ability of quickly determining the day of the week for a given date (e.g., January 17, 1893, September 4, 1943, etc.) using interesting properties in our calendar.

January the 1st AD was (as determined by calculation) a Saturday. As every simple year has 365 days, or 52 full weeks and 1 day, the year should end with the same day of the week with which it began, so the next year begins one day of the week later than the previous one. If January 1st of Year 1 was a Saturday, then the 1st of January of the second year was a day later, i.e., a Sunday, then the first day of the 3rd year is 2 days later and so on. The 1st of January of year 1923 would be 1922 (1923 - 1) days after Saturday if there were no leap years. In order to find the number of leap years, we simply need to divide 1923 by 4 (= 480). However, because of the change in calendar calculations, it is necessary to eliminate 13 days from this number. 480 − 13 = 467. To this number we must add the number of days that have elapsed since January 1, 1923 until the desired date – let's say for example, to December 14. This number is 347 days. To sum up, we need to add 1922, 467, and 347, and then divide the sum by 7, the rest of the division (which is a number between 0 and 6) will tell us which day of the week was December 14, 1923. Since this rest of the division is 6, we know that that day was a Friday:

$$
\begin{array}{r}
1922 \\
+ \quad 467 \\
\underline{347} \\
2736
\end{array}
$$

This is the general technique that is used to calculate which day in the week for any given date. In practice, things are much simpler. First of all, note that in each 28-year period, there are generally 7 leap years (which is the equivalent of a week), so every 28 years each day of the week for any date must be repeated. Furthermore, we recall that in the previous example, we have subtracted one from the year number (1923), and then in order to account for the calendar difference we have subtracted 13, i.e., we have subtracted 14 days in total, or two full weeks. Since these are entire weeks, it is clear, that they do not affect the result. Therefore, for any date in the XX century we must take into account the following: (1) the number of days that have elapsed since January 1 of the current year - in our example, 347, then (2) add the number of days corresponding to number of years that are the residue of the division of the current year (1923) by 28, and finally, (3) the number of leap years in this residue, i.e., 4. By summing the above *three numbers* (347 + 19 + 4), we obtain 370, and dividing the result by 7, we obtain the same residue 6 (Friday), which we have already find before.

Chapter VIII

$$
\begin{array}{r}
347 \\
+ \quad 19 \\
4 \\
\hline
370
\end{array}
$$

Likewise, we find that the January 15th, 1923 was a Monday (14 19 + 4 = 37; 37:7 leads to a remainder of 2). For February 9th, 1917, we would find 39 + 13 + 3 = 55, dividing 55 by 7, we get a residue 6 - Friday. For February 29 1904 we find: 59 + 0 -1 [3] = 58; the rest of its division by 7 is 2 - Monday.

Further simplification is that instead of considering the total number of days of the month (when calculating the number of days that have elapsed since January 1 in a given year), we can take only the remainder of the division of this number by 7.

Further, dividing 1900 by 28, we get a remainder of 24, which contain 5 leap years; adding them to 24 and found a total amount of 24 + 5, i.e., 29, which gives a reminder of 1 when divided by 7. Thus, we can determine that January 1st, 1900 was the 1st day of the week. Hence, for the first of each month, we get the following numbers which define the corresponding days of the week (we will call them "residual numbers").

The residual number for:

---

3 When dividing 1904 by 28, we have accounted for the fact that the year 1904 was a leap year; then we have accounted another time for the fact that February have 29 days. Therefore, it is necessary to discard the extra day.

Yakov Perelman

January 1
February   1 + 31 = 32, or 4
March   4 + 28 = 32, or 4
April   4 + 31 = 35, or 0
May   0 + 30 = 30, or 2
June   2 + 31 = 33, or 5
July   5 + 30 = 35, or 0
August   0 + 31 = 31, or 3
September 3 + 31 = 34, or 6
October   6 + 30 = 36, or 1
November   1 + 31 = 32, or 4
December   4 + 30 = 34, or 6

Remembering these numbers is not difficult. In addition, they can be recorded on the face of pocket watch, placing near each dial digit the appropriate number of points.

Now, we can determine the day of the week for any following day, for example, March 31st, 1923:

Number of days of the month 31
Residual days in March 4
Years since the beginning of the century 23
Including leap years 5
                                          Total   63

The remainder of the division of 63 by 7 is 0 – hence the day is a Saturday.

Let find day of the week for April 16th, 1948

Number of days of the month.	16
Residual days in April	0
Years since the beginning of the century	48
Including leap years	12
Sum	76

The remainder of the division by 7 is 6 – hence the day is a Friday.

Let find day of the week February 29th, 1912

Number of days of the month.	29
Residual days in February	4
Years since the beginning of the century	12
Including leap years	2
Sum	47

The remainder of the division by 7 is 5 – hence the day is a Thursday.

For dates in the preceding centuries (XIX, XVIII, etc.) you can use the same numbers, but you must remember that in the XIX century, the difference between the new and old calendar was not 13, but 12 days, in addition, the division of 1800 by 28 gives a reminder of 8, which together with two leap years in this number gives a total of 10 (or 10 - 7 = 3), i.e., the corresponding characteristic number for the dates of the XIX century is 3 - 1 = 2. So, for example, in order to calculate the day of the week for December 31st, 1864 new calendar, we first define the residual numbers, and then amend by adding 2 days:

Number of days of the month	31
Residual days in December	6
Years since the beginning of the century	64
Including leap years	16
Correction for the XIX century	2
Sum	119

The remainder of the division by 7 is 0, hence the day is a Saturday.

Find day of the week April 25th, 1886 new calendar

Number of days of the month	25
Residual days in April	0
Years since the beginning of the century	86
Including leap years	21
Correction for the XIX century	2
Sum	134

The remainder of the division by 7 is 1 - Sunday.

After these short exercises we can further simplify calculations as follow: Instead of the writing the residuals, we can write their residues modulo 7. For example, the day of the week of March 24th, 1934 can be calculated, as a result, using the following simple calculations:

Chapter VIII

Replacement for the month (24) 3

Residual days in March 4

Replacement for the numbe of years since 1900 6

Including leap years 1

Sum   0 (instead of 14)

The desired day is a Saturday.

It is this kind of simplified methods[4] that are normally used by those imaginary "math geniuses" who show their fast calculus "power" to the public. As you can see, it's all very simple and can easily be performed after a short exercise.

Calendar clock

Knowing these little secrets is not only useful when demonstrating tricks, but it is also very useful in everyday life. You can easily turn your pocket watch in a "perpetual calendar", by which we can determine the day of the week for any date of any year. For this, we need only to carefully remove the piece of glass from the clock, apply mascara on the dial[5] by adding points near the numbers as illustrated in the following figure. We already know how these points work.

---

4 There are many ways of reducing the computation of calendar dates. I have outlined here is the simplest of the known techniques, as used by German mathematician F. Ferrol, who was known for his strikingly rapid mental calculations.

5 We can use ink. However it is preferable to use mascara as it is easy to wash off the point from the dial when they are not more needed.

Yakov Perelman

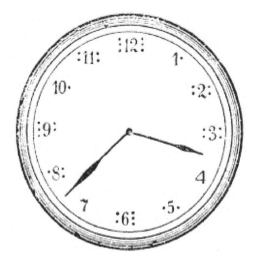

Calendar clock

This clock works only for the dates in the XX century: The number of points added to the number of months, indicate the residual numbers for each corresponding month (i.e., the reminder from the division of the number of days in the month by 7).

Then we need only to keep in mind the reminder from the division of the sum by 7 from the previous years of the century. This balance must constantly be added to the reminder (number of points) for each month up to our desired date.

In particular, for 1923, the reminder is zero because

$$(23 + \frac{24}{3}) : 7$$

For other years, it may be 1, 2, 3 ... to 7. The reminder could

Chapter VIII

be added to the number of points on the dial, as there was no need to calculate it separately. But this would make the calendar clock compatible only with the current year. It will cease to be "eternal".

It goes without saying that a "perpetual "

It goes without saying that a "perpetual calendar" of this type can be created without the need of a clock or a watch. You can simply prepare a narrow strip of paper with the appropriate residual for each month of the year as illustrated below and your ubiquitous small perpetual calendar would be ready:

I-1
II-4
III-4
IV-0
V-2
VI -5
VII-0
VIII-3
IX-6
X-1
XI-4
XII-6

CHAPTER IX
Giant Numbers

How big is a million?

The majestic impressiveness of giant numbers (million, billion, trillion, etc) has gradually faded in our eyes once paper money entered our daily lives.

If the monthly expenses of a small farmer family reach a billion rubles, and if the budget of some small institutions is expressed in trillions, then we start to think that these numbers which were once beyond our imagination are after all not that huge. Now, a number of rubles with seven digits does not allow you to buy a full pack of milk, and a billion of rubles does not allow you to buy a suit.

But it would be a great mistake to think that because of the diffusion of these numeric giants into our everyday life, we now know them better than before. A million remains for most people whatever it was: a familiar stranger. We have always been inclined to underestimate the magnitude of this number as it exceeds the power of our imagination, but when the numbers began to be expressed in millions, the value of a million shrunk in our imagination to the size of the ordinary, readily available numbers. We easily make a curious psychological error: because million rubles became a relatively small sum, we do only reduce the monetary value of this number, and we reduce the value of the number itself a well. Through habit and consistency of the ruble and the vagueness of our ideas about a million, we unconsciously continue to consider the value of the ruble as unchanged and would imagine that we finally had a chance to grasp the magnitude of millions, which turned out to be not so huge and does not deserve its alleged fame.

I heard a man innocently exclaiming when he first learned that the distance from the Earth to the Sun was 150 million kilometers:

- Only!

Another one when reading that the distance from Petrograd to Moscow was one million steps, said:

- Only *one* million steps to Moscow? But we somehow pay for the train ticket two million rubles!...

Contrary to the popular thing, our experience with money did not give us a clear idea of large numbers. Most people who worked with cash and used millions in their calculations, still do not realize clearly how these numbers are huge. Even if we perform calculations using millions of rubles, the items that are stored in our imagination have the same constant value. If you want to experience the true size of a million, try to draw a million points on a clean notebook. I do not propose to complete this work entirely (none would get patience to do that). I just propose to start it and your slow pace will make you feel what the "real" one million.

Famous English naturalist A. R. Wallace attached very great importance to developing the correct representation of a million. He suggested[1] that "in every large school creates a one room or hall whose the walls could be clearly show what a million is. For this purpose, you need to have 100 large square sheets of paper with each one measuring 4x2 feet. Now, in each sheet create and blacken quarter-inch squares and leave a space after each 10 squares in each direction, so that you have a gap around each hundred squares (10 x 10). Thus, on each sheet, you will have 10 000 black spots, easily distinguishable from the middle of the room, and if you consider all the 100 sheets, you

[1] In his book "Man's position in the universe."

Yakov Perelman

will have a million spots. Such a hall would be highly instructive especially in a nation that speak very cheeky about millions and spend them without embarrassment. Meanwhile, no one can appreciate the achievements of modern science, dealing with unimaginably large or small quantities. Imagine how great the number of one million when modern astronomy and physics have to deal with hundreds, thousands and even millions of millions.[2] In any case it is very desirable that a large town hall was arranged to indicate the magnitude of one million using such indication on the walls."

I suggest another, more affordable way for everyone to develop a clear idea of the possible magnitude of millions. To do this, you need only accept the trouble to practice mental calculations of millions and the summation of small sizes of well familiar units - steps, minutes, matches, glasses, etc. The results are often unexpected and startling.

Here are a few examples.

Million seconds

How much time would you need to count one million items assuming that you need one second per item? It turns out that if we consider an available counting time of 10 hours per day, you would end up counting for one month. To verify this approximately is not very difficult, even using mental calculus: There are 3600 seconds in an hour. 10 hours contains 36,000 seconds. In three days

2 For example, the distances between planets are measured in tens and hundreds of millions of kilometers, the distance of stars - millions of millions of kilometers, and the number of molecules in a cubic centimeter of air - million million million. - YP

you will consequently count only about 100,000 items. Counting a million items would require ten times more. To count up to a million you will need 30 days.[3] If you receive a cent for each count, at the end of the month, you would receive one million cents, or 10,000 dollars, a very decent monthly salary.

This implies, among other things, that the previously proposed work - fill up a notebook with million points - would require many weeks of very diligent and tireless work.[4] Additionally, this notebook would need a thousand pages. Nevertheless, this work was carried out once. In one circulated English magazine I recently saw playback pages of the notebook, "whose only content were neatly arranged million points, a thousand on each page." All 500 pages of the notebook were carefully filled using a pencil by the hand of one patient unprecedentedly school penmanship teacher in the middle of the last century. This work, according to the daughter of the late "author", which delivered this book to the editor, informed this one that this voluntary work took several years.

---

3 It is worth noting that a (astronomical) year contains 31,556,926 seconds.

4 The extent to which people tend to underestimate the value of millions is illustrated in the following instructive example. It is excerpt from the same Wallace who frequently warned others from belittling million: "I can arrange this room myself: it is necessary to get a hundred sheets of thick paper, line them with squares and put large black dots inside the squares. This setup would be very instructive." The venerable author, apparently believed that such work can be done by a single person. Meanwhile, we already know that it would have required an un-human amount of work – a few months of continuous operation, entirely devoted to the painstaking arrangement of black dots in large squares. Wallace made this error as he himself underestimated of the true value of a million.

Yakov Perelman

A million times thicker hair

The fineness of hair is almost proverbial. A hair is often known for its thinness. Its thickness does not exceed 0.1 mm. Imagine, however, that a hair has become a million times thicker. What would its thickness be in this case? Would it be a hand thick? A log thick? A hogshead thick? Or maybe it would reach the width of medium-sized room?

If you had never thought about such a problem and had not done the appropriate calculations, you can almost be sure that your answer will be wrong. Not only that: you will perhaps even challenge the right answer. Indeed, it turns out that a hair, whose thickness is increased a million times, would have a diameter of 100 meters! It seems incredible but go through the trouble of calculations and you will be sure that is the case: 0.1 mm × 1,000,000 = 0.1 × 1000 m = 0.1 km = 100 meters.[5]

In order to have a true idea of a million, do not confuse it with the hundreds or the thousands. To the above question of a hair, most people would think a hair, whose thickness is increased a million times, would be as thick as a barrel or a log, i.e. a value hundreds or thousands of times smaller.

Exercises with a million

We propose here a number of exercises to allow you to

[5] Here, we have done multiplication in a somewhat unusual way - instead of multiplying the number we have just replaced the unit of measure. This technique is very useful for mental calculus, and should be used in calculations in the metric system.

familiarize yourself with the value of a million. You have already seen in the previous two examples how our usual perception of a million was wrong and how it is useful to practice with this value to correct our misconception.

Let's consider the size of the well known housefly. It is about 7 millimeters in length. What would its length be after a magnification of a million times? Do not answer immediately. Multiplying 7 mm by 1,000,000, we obtain 7 kilometers - roughly the width of Moscow or Petrograd. It is hard to believe that when its length is increased a million times, a fly could cover a metropolitan city.

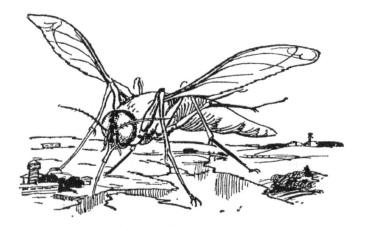

A fly after a million times magnification

If you increase the width of your pocket watch a million times, you would obtain a striking result that is hard to predict: The new watch would be 50 kilometers in width, and each digit would be an entire geographic kilometer width.

Imagine a man that is a million times taller than a

normal typical man. He would reach approximately 1700 kilometers. He would be just 8 times shorter than the diameter of the globe. In just one step, he could move from Petrograd to Moscow, and if he lies down, he would extend from Petrograd to Crimea.

Here are some final results of the same kind. The reader can perform the full calculations:

(1) Making a million steps in one direction, you could move 600 kilometers. You could move from Moscow to Petrograd.

(2) A million people lined up in single queue, shoulder to shoulder, would stretch 250 kilometers.

(3) A million glasses of water can fill a huge bowl the size of 200 barrels.

(4) A million points of typographic fonts - for example, this book - set close to each other would stretch 70-100 fathoms.

(5) Scooping up a million times a thimble, you would scoop the equivalent of a hundred buckets.

(6) A book containing a million pages would have a thickness of 25 fathoms.

(7) A million letters set close to each other would allow to have a medium forma t book containing 600 – 800 pages.

(8) A million days is equivalent to 27 centuries. Hence, since the birth of the Christ, a million days have not passed yet.

Chapter IX

Names of numeric giants

Before considering even greater numeric giants (billions, trillions, etc.) let's talk a little bit about their names. The word "million" has a universal meaning: a thousand thousands. But the words billion, trillion, etc. which were invented relatively recently have not yet received a uniform value. In financial calculations, and consequently in the everyday life we have decided to call "billions" a thousand million, and "trillion" - a million million. However, in books of astronomy and physics you find these same names but with a different meaning: a billion here does not mean a thousand million but a million million, a trillion means a million million million, a quadrillion means a million million million million, etc. In short, in astronomical books each new higher denomination is a million times higher than the precedent one. In financial calculations and everyday life, each new higher denomination is a thousand times higher than the precedent one. The table shows this difference:

	quadrillions	quadrillions	trillions	billions	millions	thousands	units
In everyday life and in financial calculations	0	000	000	000	000	000	000
In astonomy and physics	trillions		billions	millions	millions	thousands	units

You can see that what physicists call billion is called trillion by financiers, etc., so in order to avoid confusion, the name should always be accompanied by numbers. This is perhaps the only case in practice, where the amounts designated by words are rather less clear than these same amounts written using numbers. You also see that astronomers and physicists are savvier in the use of new names than financiers. One reason for this may be the fact that financiers have no reason to skimp especially that they did not have to deal with more than 12 digits numbers, while in science 20 digits numbers are frequent guests.[6]

---

[6] It should be noted, however, that conventional numeric designations of very large numbers are used only in popular-science books. In scientific books about physics and astronomy, the commonly used notations are 10^{12} for a billion, 10^{18} for a trillion, 27×10^{15} for twenty-seven thousand billion, etc. This notation saves place, and allows to easily perform various operations on numbers.

A billion

The word "billion" is used in the sense of a thousand million in monetary calculations as well as in hard sciences. But, in old days, in some countries such as Germany and in America, the word Billion was sometimes used to refer to a hundred million (not a thousand million). This explains the use of the word "Billionaire" at these old times, as nobody, even the riches, could reach fortunes of billion (in the sense of thousand million) dollars. Rockefeller huge fortune was estimated shortly before the war to be at "only" $900 million, and other billionaires had ever smaller fortunes. It is only during the war, that real billionaires (in our sense of the word) appeared in America.

To obtain an idea of the size of a billion, think about the book you are reading now. This one has a little more than 200,000 letters. In five equivalent books, you would have one million letters. A billion letters would require a stack of 5,000 copies of this book. If this stack was neatly folded, it would be as tall as a pillar in St. Isaac's Cathedral.

In order for a clock to beat one billion seconds, it will need more than 30 years. A billion minutes is equal to more than 19 centuries. Mankind has started counting the second billion of minutes of our calendar only twenty years ago (April 29, 1902 at 10 h 40 min).

A billion and a trillion

Feeling the enormity of these numeric giants is difficult even for a person who is experienced in dealing with millions. Giant million is a dwarf when compared with a billion which is the next unit after the million. Usually, we fail to realize the big difference between a million, billions and trillions. Here we are like those primitive peoples who can count only up to 2 or 3, and all the numbers above these ones are equal and are referred to using the word "a lot". "Just how insignificant seems the difference between two and three - says-known German mathematician professor G. Schubert - similarly in several modern popular cultures the difference between billions and trillions seems insignificant. At least they do not think about that one of these numbers is a million times bigger than the other and consequently if the first referred to the distance between Berlin and San Francisco, the second would refer to the width of the street."

The relationship between millions, billions and trillions can be given some clarity as follow: In Petrograd, there are about one million inhabitants now (1923). Now imagine that we have a straight line of cities similar to Petrograd. Imagine that we have a million of them. Such a line would extend 7 million kilometers (20 times the distance between the Earth and the Moon) and would house a billion of inhabitants... Now imagine that we don't have a straight line of a million of such cities, but instead we have a square full of such cities which side built using a million of Petrograds. In this square you would have a trillion inhabitants... If we had one trillion bricks, we could use them to build a dense layer on the solid surface of the Earth. This layer would raise to the height of four-story building (8 years).

If all the stars that are visible in the strongest telescopes were celestial globes, i.e., at least 30 million stars – and were inhabited and populated each one with 20 times more people than our earth – they would not, even combined, be

Chapter IX

184

able to host a total one trillion people.

Finally, the last image is borrowed from the world of tiny particles that make up every natural body - from the world of molecules. A molecule is so small that a point printed on this book has the width of a million times that of a molecule. After all the preceding exercises, you can already have a certain idea about the smallness of a molecule. Now imagine a trillion of these molecules,[7] closely strung on a thread. How long would this thread be? It could wrap the globe at the equator seven times!

Quadrillion

In the old days (XVIII century) a table was contained in Magnitsky's "Arithmetic" which we have repeatedly mentioned in this book. The one included the names of different classes of numbers up to the quadrillion, i.e. one unit with 24 zeros.[8]

It should be noted that the human imagination has limits and his mind could not imagine unusually giant numbers. Additionally the set of numbers presented in Magnitsky's are sufficient for the calculation of all visible things in this world. This is why in most of the case, we don't need to consider numbers beyond the quadrillion.

It is also interesting to note that Magnitsky was a seer in this case. Modern science does not have the need for

7 Each cubic centimeter of air (i.e. approximately in a thimble) there is a total of between 20 and 30 trillion of molecules. You wonder which aspect should marvel you more: the huge number of molecules or their incredible smallness...

8 Magnitsky adhered to the classification of numbers which gives each new name to a million of lower name units (e.g. a billion is million millions, etc.)

Yakov Perelman

higher numbers that quadrillions.

The distances between the farthest star clusters are estimated (according to the latest estimates of astronomers) at around 200,000 "light years"[9]. These distances are these of stars that can be visible using the strongest available telescopes on Earth. Distance between stars located "inside the Sky" is, of course, smaller. The total number of stars has been calculated to be "only" hundreds of millions. In term of age, the oldest stars do not exceed, at the most generous estimate, a billion years. The weight of some stars is estimated to reach the thousand quadrillion tons.

Turning to the opposite side, to the world of very small quantities, and here we do not feel the need to use far more than the quadrillion. The number of molecules in a cubic centimeter of gas - one of the biggest numbers in the realm of small quantities - is expressed in tens of trillions. The number of oscillations per second for the fastest oscillating waves of radiant energy (X-rays) does not exceed 40 trillion. The field magnitude of the smallest object that exists in nature - atoms of positive electricity - is not lower than a trillionth of a millimeter.

If we wanted to count how many buckets of water are embodied in all the oceans of the Earth, we would not reach a quadrillion, because the total volume of the 1440 million cubic kilometers of water that are contained in the oceans would lead "only" to 1440 trillion liters or 120 trillion buckets. To calculate the number of drops in the oceans (assuming a drop of water has a volume of 1 cubic millimeter - which is quite small), we do not need to use magnitudes beyond the quadrillion, because this leads only

9 A light year is the distance traveled by a beam of light in one year (light travels 300,000 kilometers in a second).

to 1440 quadrillion. Magnitsky was correct and wise when he said that the quadrillion dominate all the numbers.

Cubic geographical mile and cubic kilometer

We will consider now an arithmetic (or rather, perhaps, a geometric) giant of a special kind - on a cubic geographical mile, we are referring here to a geographical mile[10] - 7 versts long[11], and not to standard cubic mile (a cubic geographical mile is approximately 97.97 standard cubic mile). Our imagination copes pretty poorly with cubic measures and we usually underestimate significantly their value - especially for large cubic units, which we have to deal with in astronomy. If we cannot imagine properly the volume of a cubic geographical mile, how can we imagine the extent of the Earth, the planets, and the sun? We should therefore devote some time and attention to try to have a proper and clear view of a cubic geographical mile.

For this purpose, we will borrow some picturesque settings from the talented German popular science writer A. Bernstein, but in a slightly altered form. Here is a long extract from his rare little book - "A fantastic journey through the universe" (which appeared more than half a century ago).

"When we are facing a direct road, we can see a geographical mile ahead. Now place a mast one geographical mile long at one end of the road at a distance of a geographical mile

10 The geographical mile is unit of length determined by is 4 minutes of arc along the Earth's equator. and is defined as approximately 7421.5 meters (4.61 miles approximately).

11 A verst is an old Russian unit of length. One verst is equal to 1.0668 km or 3,500 English feet.

Yakov Perelman

from us. Look up and observe the height of the mast. Now suppose that next to this mast stands a human statue with a nearly identical height. The statue would be seven versts of high. In such statue, the knee would be 900 fathoms high.[12] It would be necessary to stack up 18 St. Isaac's just to reach the knee of such a monument. 25 Egyptian pyramids would need to be piled up one on another to reach the waist of the statue.

Imagine now that we have installed two of these masts one geographical mile away from each other, and connected both masts by the panels. We would obtain a wall that is a geographical mile high and a geographical mile long. It would have an area of one square geographical mile.

If there were really such a wall, for example, along the Neva River in St. Petersburg, then the climatic conditions of this fabulous city would have significantly changed: the north side of the city could see more severe winter, while the southern side would have enjoyed an early summer. During the month of March, on one side of the wall, you would enjoy a tour on a boat, while on the other side, you would be having a ride on a sleigh or skiing on ice... but we digress from numbers here...

We have a wooden wall, standing vertically. Imagine another four such walls, put together as a box. Cover it with a lid that is a geographical mile long and a geographical mile wide. The volume of this box is one cubic geographical mile. Let us now see how huge the box is, i.e., what and how much can fit in it.

Let's remove the lid, and throw all the buildings of Petrograd in the box. These ones would take up very little

12 A fathom is an old unit of length and is equal to 6 feet or 1.8288 meters.

space there. Add Moscow buildings, and all the buildings in the provinces. These ones would cover only the bottom of the box. So we need to look for materials somewhere else. Take Paris with all its triumphal gates, towers and threw them in the box. All this barely changes the level in the box. Add London, Vienna, and Berlin. But since all this is not enough to fill the box, we start to indiscriminately throwing in all the cities, forts, castles, villages, and individual buildings. Still not enough? Threw everything made by human hands in Europe, but this will only get the box filled to the quarter. We add all the ships of the world, but this does not help. Add all the Egyptian pyramids, all the rails of the Old and New Worlds, all the trains, cars and factories of the world - all what has been made by people in Asia, Africa, America and Australia. The box will barely get filled to its half. Now let's try whether you can fill it with people.

Collect all the straw and all the cotton paper that exists in the world and spread it in a box - we get a layer that protects people from injuries associated with this experiment. All the German population - 50 million people - will settle in the first layer. We cover them with a foot thick soft layer of straw and stack another 50 million of people. We repeat the same process and add in the entire population of Europe, Asia, Africa, America, and Australia. These ones would barely fill in 35 layers. Assuming a 1 meter per layer, 50 times the earth population would be needed to fill the second half of the box.

What do we do? If we wanted to put in a box all the wildlife - all horses, oxen, asses, mules, sheep, camels, and impose on them all the birds, fish, snakes, anything that flies or crawls, we would not be able to fill the box up to the top without using rocks and stones.

Yakov Perelman

We are talking here about one box of one cubic geographical mile and what it could contain. Right, you can give it some respect!

A cubic geographical mile

Is it possible that one cubic geographical mile is so huge? Why is this box so difficult to fill? Can't we come up with a machine that would make enough material to fill it? Let's think about an idea like this: Let's construct a brick factory and set up a machine that would prepare one brick cube of 1 foot every second. Set it up so that it would operate day and night without interruption and arrange every brick made in the box. Let's start the machine. Our eyes are barely able to follow the work. Wait: probably, the machine will complete its work soon.

Indeed, soon... We can precisely calculate this. The machine produces one brick every second. That would be 60 bricks per minute, 3600 per hour, 86,400 day, and about 31 million per year.

But how many of these bricks are needed to fill the box? A square with sides 7 versts, or 24,500 feet, will have a volume of round 600 million square feet. Therefore 600 million bricks need to be laid down in the first layer of the box. And as the factory produces annually a total of 31 million bricks, it is clear that to cover only the bottom of the box you will have to wait for about 20 years.

The box also has a mile in height, which means that the machine needs to fill 24,500 of these layers that fill the bottom. Doing the multiplication, we see that our machine will not finish its work so soon, it initially seemed to us. It must operate night and day for almost half a million years to complete its task...

Such is a cubic geographical mile. But how about our Earth which is made of 660 million of such boxes! If the cubic mile deserves some respect, our Earth deserves even more respect.

Now that incredibly huge cubic geographical mile (about 350 cubic kilometers) can be felt by the reader. We can add that if we have to fill a cubic geographical mile with wheat grains, we would need a few trillions of these ones.

The capacity of a cubic kilometer is impressive as well. It is easy to calculate, for example, that a cubic kilometer box could accommodate 5,000 billion closely packed matches, and in order to make so many matches, a factory that produces one million matches per day, would have to work 14 million years. In order to deliver this number of matches, 10 million carts would be needed. These ones could be linked to a train having the length of 100,000 kilometers, i.e., 2 ½ times the earth's equator:

Yakov Perelman

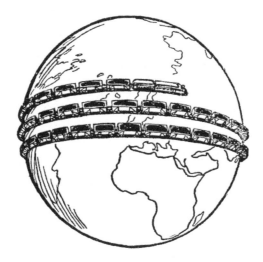

After what has been said about the cubic geographical mile, gigantic numbers, such as the trillion and the quadrillion, grow further in our minds.

Chapter IX

CHAPTER X
Lilliputian Numbers

In his wanderings, Gulliver found himself among the Giants after leaving the Lilliputians. We have travelled in the opposite direction: After getting acquainted with numeric Giants, we are going to the world of Lilliputians, to extremely small numbers.

There is no difficulty in finding representatives of this world: one simply needs to consider the inverses of a series of increasing numbers, millions, billions, trillions, etc., i.e., divide one by these numbers. The resulting fractions

$$\frac{1}{1\ 000\ 000}, \frac{1}{1\ 000\ 000\ 000}, \frac{1}{1\ 000\ 000\ 000\ 000} \quad \text{ETC}$$

are typical Lilliputian numbers, as they are several times smaller than one. They a million, billion, trillion times smaller than one.

We can see that for each giant number, a dwarf one can be found, and consequently the numbers of dwarf numbers is equal to the number of giant numbers. We also use shorthand notations to write them down. In a very large number of scientific books (astronomy, physics), they are denoted as follow:

1,000,000	10^6
10,000,000	10^7
400,000,000	4×10^8

etc.

Accordingly, the Lilliputian numbers are designated as follow:

Chapter X

$$\frac{1}{1\ 000\ 000} \cdots\cdots\cdots\cdots\cdots\cdots\cdots 10^{-6}$$

$$\frac{1}{1\ 000\ 000\ 000} \cdots\cdots\cdots\cdots\cdots\cdots 10^{-9}$$

$$\frac{1}{1\ 000\ 000\ 000\ 000} \cdots\cdots\cdots\cdots\cdots 10^{-12} \quad \text{ETC}$$

However, is there a real need for these small fractions? Do have a need to deal with such small quantities? It is interesting to talk more about this topic.

Lilliputian time

In normal life, a second is such a small amount of time that a fraction of a second cannot be noticed under any circumstances. What might happen, for example, in one-thousandth of a second? It is easy to write: $1/1000^{th}$ of a second, but this is purely paper value, because nothing can happen in such an insignificant amount of time. But this is just an impression.

A train traveling at 36 kilometers per hour makes 10 meters per second, and consequently, during a $1/1,000^{th}$ of a second time, it would move one centimeter. Sound in the air travels 33 centimeters (about a foot) during a $1/1,000^{th}$ of a second, and a bullet leaving a gun barrel at the speed of 700-800 meters per second, travels over the same period of time one whole arshin.[1] The Earth travels 30 meters

[1] A Russian measure unit. An arshin is exactly equal to twenty-eight English inches (71.12 cm).

Yakov Perelman

every $1/1,000^{th}$ of a second, in its revolution around the Sun. A string making high tone sounds makes 2-4 and more complete oscillations in a $1/1,000^{th}$ of a second. Even a mosquito has time to flap it wings upward or downward during this time. Lightning lasts far less than $1,000^{th}$ of a second, i.e., during this period of time, it has time to appear and disappear while crossing distances of several versts.

But - you will object – $1/1000^{th}$ of a second is not a genuine dwarf, as no one would call a thousand a numeric giant. If you consider a millionth of a second, you can probably argue it is. You can argue that this is an interval of time during which nothing can happen. But this is also wrong. Even one millionth of a second is not an overly small period of time for modern physics, for example. Light and electric phenomena (in physics) very often deal with much smaller fractions of a second. First, we can remember that light travels 300,000 kilometers per second (in a vacuum), hence in a millionth of a second, a light beam travels a distance of 300 meters – about the same distance crossed by sound in a whole second.

Then, light is a wavelike phenomenon, and the number of light waves that weave through a point in space every second can reach hundreds of billions. Light waves which cause a sensation of red light when acting on our eyes have an oscillation frequency of 400 billion per second. This means that within one millionth of a second 400,000,000 waves enter in our eyes, or equivalent, one wave enters our eyes every $1/400,000,000,000,000^{th}$ of a second. That is a genuine numeric dwarf!

But without doubt, one millionth of a second is still a true giant when compared with smaller Lilliputians. We

Chapter X

can find these ones when studying the physics of X-rays. These remarkable rays have a wonderful property as they can penetrate many opaque bodies. These rays are similar to visible rays as they are a wave phenomenon, but their oscillation frequency is considerably greater than that of visible rays. This frequency reaches 25,000 billion per second. An X wave is followed immediately by another one 60 times more often than with the rays of visible red light. Lilliput Gulliver was only 10 times bigger than Lilliputians and yet he seemed like a giant. Here one Lilliputian is five dozen times bigger than the other. Consequently, it has all rights to be named a giant when compared to it.

Lilliputian space

It is interesting to consider now the shortest distances that can be measured and evaluated by modern researchers.

In the metric system the smallest unit of length for common people is the millimeter. This unit is about half the thickness of a match. To measure objects that are visible to the naked eye, such a unit of length is shallow enough. But to measure ciliates, bacteria and other small objects that are distinguishable only through a strong microscope, the millimeter too burly. In order to address this situation, scientists, use other smaller units such as the micron, which is 1,000 times smaller than a millimeter. The so-called red cells which are present in blood with a concentration of tens of millions in every drop of our blood have a length of 7 microns and a thickness of 2 microns. A stack of 1,000 such cells has the thickness of a match.

No matter how shallow it seems to us, a micron is still

Yakov Perelman

too large for some distances that have to be measured in modern physics. Tiny particles that are known as molecules and that make up the substance of all natural bodies are invisible even by a microscope. They are composed of even smaller atoms whose size is $1/10,000^{th}$ to $1/1,000^{th}$ of a micron. So if you consider a 1 millimeter speck of dust, it turns out that in the best scenario an atom is a millionth of this speck (and we already know how great a million!). Aligning 1 million of these atoms one after another in a straight line, you would obtain one millimeter.

To vividly imagine the extreme smallness of atoms, let consider the following image: Imagine that all the items on the Earth have increased in size a million times. The Eiffel Tower (300 meters high) would be 300,000 kilometers high and its tip would go into space and be in the near vicinity of the Moon's orbit. People would be 1800 km high, one step of this giant men would carry them 600 to 700 versts. The billions of tiny blood cells that are floating in our blood would be 7 meters in diameter each. Hair would have a thickness of 117 meters. A mouse would reach 100 versts in length, and a fly would be 8 versts long. How big would an atom in this new world? You wouldn't believe: An atom would be the size of a typographical point... from this book!

With the atom, we reach the extreme limits of spatial smallness. The measurement of the dimensions of this one requires sophisticated physics techniques. But now, we know that even the atom, which we thought was the indivisible building block of the universe, consists of much smaller pieces and is the scene of the action of powerful forces.

An atom, for example a hydrogen one consists of a central

Chapter X

core and a fast electron orbiting it. Without going into further details, let's talk only about the size of these atom components. The diameter of the electron is measured billionth of a millimeter, while that of the nucleus is a thousand billionth of a millimeter. In other words, the diameter of the electron is almost a million times smaller than that of the atom, while the diameter of the nucleus is a billion times smaller than that of an atom. If you wish to compare the size of the electron with the size of a speck of dust, the calculation will show you that the electron is the speck of dust what this speck of dust is to the globe!

You can clearly see that if an atom is a Lilliputian when compared to usual objects, it is at the same time, a giant when compared to the electron within its composition with the same scale when comparing the entire solar system to the Earth.

We can now compose the following instructive series in which each member is a giant when compared to the preceding one, and a dwarf when compared to the following one:

The electron,
the atom,
the speck,
the house,
the Earth,
the Solar System,
the distance to the North Star.

Each member of the series is about a quarter of a million times[2] bigger than the preceding one and the same

---

2   This refers to the linear dimensions, that is, the diameter of the atom, the diameter of the solar system, the height or length of the house, etc.

Yakov Perelman

factor smaller than the following one. Nothing proves so eloquently the relativity of the concepts of "big" and "small" more than the above series. In nature, big or small are definitely subjective. Everything can be overwhelmingly huge and vanishingly small, depending on how you look at it, and compare it to something else. We will finish with the words of a British physicist[3]: "Time and space are concepts that are purely relative. If today at midnight all the objects - including ourselves – are reduced in size by a factor of 1000, we would absolutely not have noticed this change. There would be no indication that there has been such a reduction. Similarly, if all the events and all the clocks have sped up in the same respect, we, likewise, did not suspect anything would change about this. "

Giants and Dwarfs

Our presentation of the giants and the dwarfs of the world of numbers would be incomplete if we did not tell the reader about one amazing curiosity of this kind - a gimmick, though not new, but worth a dozen new gimmicks. To understand this one, let's with the following, seemingly very unpretentious question:

What's the biggest number you can write with three digits?

We need to answer: 999 - but you probably already suspect that the answer is different, otherwise the task would be too simple. And indeed, the correct answer is written as:

$$9^{9^9}$$

3 Fournier d'Albe. "Two new world". 1907

This expression means "nine power nine power nine." In other words: you have to make a product of a huge number of nines. Let start by multiplying nines 9s:

$$9 \times 9 \times 9 \times 9 \times 9 \times 9 \times 9 \times 9 \times 9.$$

Just start this calculation to feel the enormity of the upcoming result. If you have enough patience to perform the multiplication of nine nines, you will get the following number:

$$387,420,489.$$

The homework has just started: Now we need to find the value of $9^{387420489}$, i.e., the product of 387,420,489 nines. Basically, you have the make around 400 million multiplications... You will certainly not have time to complete all these calculations. But I am unable to tell you the exact final result. There are three respectful reasons for that. First, nobody has ever calculated this number exactly. Only the approximate result is known. Second, even if this number was calculated, then, printing it would require at least a thousand books like this one, because our number consists of 369,693,100 digits; using an ordinary font size, it length would be around 1000 versts... Finally, even if I were supplied a sufficient amount of paper, I would not be able to satisfy your curiosity and now you will understand why: If I was able to write without interruption at the pace of two digits per second, I would write down 7,200 digits per hour, and if I were able to write these digits continuously day and night without interruption, I would write 172,800 digits per day. Therefore, if I continue writing these digits around the clock day and night without taking any holidays, I would need at least seven years in order to complete this number...

Yakov Perelman

You see that the number of digits in our result is unimaginably huge. As huge as it may be, can we express it using some known huge numbers?

Archimedes once calculated the number of grains of sand that would be necessary if all the fixed stars were filled with this matter. He got a number with 64 digits. But our number is not composed of 64 digits, it is composed of 370 million digits, hence it is immeasurably bigger than the huge number of Archimedes.

Let proceed in the same way as Archimedes', but instead of calculating the number of sand grains, let calculate the number of electrons. Let consider the dimensions of the universe and take the highest limit value acceptable by modern science. Namely, there are reasons to believe that the diameter of the universe cannot exceed the distance traversed by a light ray in a billion of years (light travels 300,000 kilometers per second). Imagine now that the whole universe with the above size is completely filled with a dense metal - platinum, in which each atom contains 78 electrons. How many electrons would this virtual universe contain? Calculations give a result consisting of "just" 100 digits. How many "platinum universes" would be needed to accommodate

$$9^{9^9}$$

electrons? The required number of such universes would itself contain about 369,693 digits... You can see that filling the whole universe - the greatest space we know – with electrons, i.e., the smallest object we know – will not allow us to reach the value of our gigantic number (not even a small part of it), which is modestly hiding under the writing:

$$9^{9^9}$$

Our attempt to get acquainted with this masked giant has only led to confusion. If you were asked what is the smallest number you can write with three digits you will not be satisfied with numbers like these:

$$100; \quad \frac{1}{99}; \quad 0,01,$$

You will probably write something like:

$$\frac{1}{9^9}.$$

This is indeed a very small number, because it is equal to

$$\frac{1}{387420489}.$$

However, modest intrusions into algebra give you the means to write a much smaller number, namely

$$9^{-9^9}.$$

This number is equal to:

Yakov Perelman

$$9^{-387420489}, \text{ i.e. } \frac{1}{9^{387420489}}$$

In order words, we have here a familiar giant number, but since it is located in the denominator, it becomes a Lilliputian.

CHAPTER XI
Arithmetic Travel

Your trip around the world

Fifteen years ago, I met a person in a social event. When we exchanged our business cards, the designation of this occupation was quite extraordinary. His title read: *"The first Russian around the Earth foot traveler."* At the time, I have seen Russians travelling to other parts of the world and even completing trips around the world, but I have never seen a traveler completing a trip around the world "on foot". Curious, I hurried to get acquainted with this adventurous and indefatigable man.

This wonderful traveler was young and has a very modest appearance. Asked when he had made his extraordinary journey, he explained to me that he is doing it right now. Which route did he take? Shuvalovo[1] - Petrograd, the further he explained to me, the more it became clear that his title of the *"the first Russian..."* was rather dubious as he has never crossed the borders of Russia.

- Then, how are you going to complete your journey around the world? I asked.
- The main objective is to reach a distance equivalent to the circumference of the Earth, and this can be done in Russia without the need to cross its borders. To my bewilderment, he added: I have already completed ten kilometers and remains...
- A total of 39,990. Godspeed!

I believe he will be able to complete that distance. I have no doubt about that. Even if he did not travel anymore, and

---

1 Shuvalovo – A small station 10 kilometers from Petrograd.

immediately returned to his native Shuvalovo hopeless and lived there forever - he would be able to travel 40,000 kilometers. I am afraid that he is not the first and not the only person who has made such a feat. Me, you and most other Russians have the same right to the title "*Russian around the Earth foot traveler*," as per the above walker understanding. Because each of us, even if he stayed his entire life in his hometown, had during his life, without knowing it, walked this distance, and even distances that are longer than the circumference of the Earth. Now, small arithmetic calculation will convince you of that.

Every day, it is likely that you spend at least 5 hours on your feet: you go through the rooms, the garden, the streets, in a word, you move. If you have a pedometer in your pocket (a device for counting steps) it would show that you do everyday at least 30,000 steps. But even without a pedometer is clear that the distance you travel in the day is very impressive. The slowest walking man can do 4-5 kilometers in an hour. During a day, i.e., in 5 hours, he would do 20-25 kilometers. Now it remains to multiply this distance by 360 in order to estimate the distance done in one year:

$$20 \times 360 = 7200, \text{ or } 25 \times 360 = 9000.$$

Thus, the most sedentary people, perhaps these who have never left their hometown, travel about 8,000 kilometers on foot annually. As the circumference of the Globe is 40,000 kilometers, it is easy to calculate how many years are needed to walk a distance equal to this circumference:

$$40000 / 8000 = 5.$$

So, in 5 years you do a distance that is equal to the

circumference of the globe. Every 13-year-old boy, if we assume that he started walking after two years of age - has made two trips "around the world". Every 25-year-old man has completed least 4 such trips. And before a man reaches 60 years, he would have gone ten times around the globe, i.e., crossed a distance that is longer than the distance from the Earth to the Moon (380,000 kilometers). Such an unexpected result is impressive when we know that we are just walking inside our homes and around them.

Your ascent of the Mont Blanc

Here's another interesting calculation. If you ask the postman who everyday delivers letters to addressees, or a doctor, visiting his patients during a busy day, whether they made the ascent of Mont Blanc - they certainly would be surprised by such strange question. Meanwhile, you can easily prove to each of them, that, even if they are not climbers, they have probably climbed during their years of service a height that is greater than even the greatest peaks of the Alps. One has only to calculate how many stairs do a postman or a doctor climb every day, when delivering letters or visiting patients. It turns out that the most humble postman, or doctor, who has never even thought about sports, may score world records in term of climbed distances.

For the sake of illustration, let take rather modest average figures; assume a doctor visit ten people every day on average. These people may live on the second, third, fourth, fifth floors etc - let take the third floor as an average. The height of the third floor is 10 meters: hence a doctor climbs 100 meters on a daily basis. The height of the Mont Blanc is 4800 meters. Dividing it by 100, you will find that our humble doctor performs an ascent of the Mont Blanc every 48 days...

A postman in his "Mont Blanc ascend"

Every 48 days or about 8 times a year, a postman or a doctor climb stairs to a height equal to the highest peak in

Europe. Tell me how many athletes climb the Mont Blanc 8 times per year?

It is not necessarily to be a postman, to perform such feats. Most people without knowing it do it. I live in an apartment on the 2nd floor. A staircase with 20 stairs is needed to reach the apartment. This number is seemingly very modest. However, on a daily basis, I use these stairs 5 times. Additionally, I visit a friend, who lives on the 4th floor almost on a daily basis. Consequently in average, we can assume that every day, I go up 20 stairs 7 times. This is 140 stairs per day. How does this make in a year?

$$140 \times 360 = 50400.$$

So, every year I go up more than 50,000 stairs. If I'm going to live up to the age of 60, I'd have climbed to the top of an imaginary ladder with three million steps. What a gigantic height climb! But this is nothing when compared with the heights climbed by some people by virtue of their profession, such as for example elevators workers. Someone has calculated that, for example, an elevator worker of a New York skyscraper would reach in 15 years of service the height of the moon...!

Traveling plowmen

Take a look at the strange pattern in the following figure. Who are these heroic Plowmen who furrow round the globe?

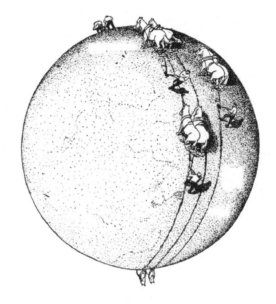

"Who are those heroic Plowmen who furrow round the globe?"

Do you think this drawing was created by an artist having a too raging and uncontrollable imagination? Not at all: the artist has simply depicted graphically what reliable arithmetic calculations have produced. Each Plowman and his plow cross over several years (4 -6 years) a distance which is equal to the circumference of the globe. This unexpected result can be obtained by arithmetic calculations that can be easily carried out by the reader.

Inconspicuous journey to the ocean floor

Very impressive travels are carried out by people working in basements and underground warehouses, etc. Several times a day, they take the stairs to move objects and other merchandise. Within a few months they cross the distance of several kilometers. It is easy to calculate how much time is needed for a basement worker to go down a distance equal to the depth of the ocean. If the stairs' height is, say, only 1 fathom, i.e. 2 meters, and the worker use them let's say 10 times every day, then in one month, he would descend a depth of $30 \times 20 = 600$ meters, and in a year a depth of $600 \times 12 = 7200$ meters, i.e., more than 7 kilometers. Remember that deepest place on the continents is only 2 kilometers deep!

Let's suppose the stairs lead to the bottom of the ocean, then any basement worker would have reached the bottom of the ocean in about a year (the maximum depth of the Pacific Ocean is about 9 kilometers). Without knowing it, a worker would have descended the depth of the Pacific Ocean and would have experienced the mysterious realm of bizarre deep-sea creatures, which until now, are seen only by researchers exploring deep seas.

Travelers sitting still

Do not think that arithmetic travel is applicable to moving people only. There are people who, sitting motionless at their work, make long pilgrimages nevertheless. How far would a weaver sitting on his seat working diligently on his sewing machine travel? It turns out he does not escape the fate of being a globe traveler. Every second, his nimble fingers manage to move back and forth about 50 centimeters. How much would they do in an hour?

$$50 \times 60 \times 60 \text{ cm} = 180\ 000 \text{ cm} = 1800 \text{ meters}.$$

Thus, the weaver travels almost 2 kilometers per hour. In an 8-hour working day, he would travel more than 14 kilometers.

It is easy to calculate how long it would take him in order to complete a full revolution around the Earth. Dividing the circumference of the Earth (40,000 kilometers) by 14, we get more than 2800 days. This means that after 8 years of hard work, our weaver would had made a full trip around the world with the ends of his fingers.

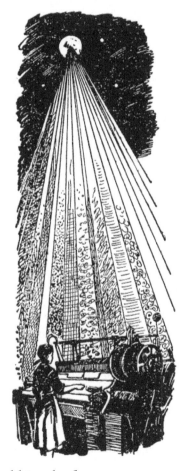

How long would it take for a weaver to reach the Moon?

Yakov Perelman

In fact, you would not be able to find a man who somehow did not make a full trip around the globe. Today, an extraordinary person is not the one who did travel around the world. Instead he is the one who did not. And if anyone would assure you that he has not made such a feat, I hope you can now "mathematically" prove to him that he is an exception to the general rule.

- The End -

ISBN : 9782917260388

People who read this book, also read: Physics for Entertainment (volumes 1 and 2) by Yakov Perelman:

Discover them on Amazon.com